COMPETITIVE INDUSTRIAL DEVELOPMENT IN THE AGE OF INFORMATION

As the global economy has evolved and matured, transnational corporations, small- to medium-sized enterprises (SMEs) and governments have emerged as the principal players in industrial development. Knowledge and technology have also taken on new importance as driving forces for economic growth. This work examines policy issues arising from these developments, which will be of fundamental importance in a twenty-first century Age of Information. Discussion focuses upon:

- the role of national and subnational industrial policy in promoting economic performance
- the impact of tax policy upon domestic technology development
- bank financing of SMEs
- the need for cooperation in the development of economically viable industrial policy

With special reference to the technology sector, contributors demonstrate that globally oriented industrial policy must promote both competition and cooperation in the domestic economy. Domestic competition must be viewed as a function of global competition, strengthening a country's technological and economic development. However, domestic cooperation allows parties to compete on a global scale. It is concluded that this 'New Competition', embracing cooperation, is the key to understanding the four major policy issues discussed.

Competitive Industrial Development in the Age of Information draws upon detailed case studies of technology sector industries to provide theoretical and empirical analyses of changing economies and changing policy needs. With contributors from academic, legal, financial and policy advisory backgrounds, this work advances research into policy questions of increasing importance.

Richard J. Braudo is a Toronto- and Dallas-based Business Economist and Management Consultant. He has been a Business Economist for more than fifteen years and a lawyer since 1996. His research includes work on competition and constitutional law, health care, international trade policy, and e-commerce.

Jeffrey G. MacIntosh is a Professor of Law at the University of Toronto. He has written extensively on topics relating to corporate and securities law and has recently focused his research upon the financing of small, high technology companies.

ROUTLEDGE STUDIES IN INTERNATIONAL BUSINESS AND THE WORLD ECONOMY

1 STATES AND FIRMS
Multinational enterprises in institutional competition
Razeen Sally

2 MULTINATIONAL RESTRUCTURING, INTERNATIONALIZATION AND SMALL ECONOMIES
The Swedish case
Thomas Andersson, Torbjörn Fredriksson and Roger Svensson

3 FOREIGN DIRECT INVESTMENT AND GOVERNMENTS
Catalysts for economic restructuring
Edited by John H. Dunning and Rajneesh Narula

4 MULTINATIONAL INVESTMENT AND ECONOMIC STRUCTURE
Globalization and competitiveness
Rajneesh Narula

5 ENTREPRENEURSHIP IN A GLOBAL CONTEXT
Edited by Sue Birley and Ian Macmillan

6 THE GLOBAL STRUCTURE OF FINANCIAL MARKETS
An overview
Edited by Dilip K. Ghosh and Edgar Ortiz

7 ALLIANCE CAPITALISM AND GLOBAL BUSINESS
John H. Dunning

8 MULTINATIONAL ENTERPRISES FROM THE NETHERLANDS
Edited by Roger van Hoesel and Rajneesh Narula

9 COMPETITION, GROWTH STRATEGIES AND THE GLOBALIZATION OF SERVICES
Real estate advisory services in Japan, Europe and the United States
Terrence LaPier

10 EUROPEAN INTEGRATION AND FOREIGN DIRECT INVESTMENT IN THE EU
The case of the Korean consumer electronics industry
Sang Hyup Shin

11 NEW MULTINATIONAL ENTERPRISES FROM KOREA AND TAIWAN
Beyond Export-led Growth
Roger van Hoesel

12 COMPETITIVE INDUSTRIAL DEVELOPMENT IN THE AGE OF INFORMATION
The role of cooperation in the technology sector
Edited by Richard J. Braudo and Jeffrey G. MacIntosh

13 THE GLOBAL RESTRUCTURING OF THE STEEL INDUSTRY
Innovations, Institutions and Industrial Change
Anthony P. D'Costa

14 PRIVATISATION AND LIBERALISATION IN EUROPEAN TELECOMMUNICATIONS
Comparing Britain, The Netherlands and France
Willem Hulsink

15 THIRD WORLD MULTINATIONALS
Paz Estrella Tolentino

COMPETITIVE INDUSTRIAL DEVELOPMENT IN THE AGE OF INFORMATION

The role of cooperation in the technology sector

Edited by Richard J. Braudo and Jeffrey G. MacIntosh

Routledge
Taylor & Francis Group
LONDON AND NEW YORK

First published 1999 by Routledge
2 Park Square, Milton Park, Abingdon, Oxon OX14 4RN
605 Third Avenue, New York, NY 10017

Routledge is an imprint of the Taylor & Francis Group, an informa business

Typeset in Sabon by
BC Typesetting, Bristol

British Library Cataloguing in Publication Data
A catalogue record for this book is available from the British Library

Library of Congress Cataloging in Publication Data
Competitive industrial development in the age of information: the role of cooperation in the technology sector/edited by Richard J. Braudo and Jeffrey G. MacIntosh.
p. cm.
Based on a May 1995 workshop held at the Faculty of Law, University of Toronto, and sponsored by the Faculty of Law with the law firms of Borden & Elliot and Stikeman, Elliot; and the Canadian Bar Association-National Competition Law Section.
Includes bibliographical references and index.
1. High technology industries – Government policy – Congresses. 2. Industrial policy – Congresses. 3. Competition – Government policy – Congresses. 4. Corporations – Taxation – Law and legislation-Congresses. 5. Industrial laws and legislation – Congresses. I. Braudo, Richard J., 1954– . II. MacIntosh, Jeffrey G.
HC79.H53C656 1999
338.4′762–dc21 98-26660
CIP

ISBN 13: 978-0-415-17854-9 (hbk)

CONTENTS

LIST OF FIGURES AND TABLES

FIGURES

TABLES

LIST OF CONTRIBUTORS

Richard J. Braudo, Business Economist and Lawyer, Toronto.

Paul Halpern, Faculty of Management, University of Toronto.

Robert Howse, Faculty of Law, University of Toronto.

Derek Ireland, Office of Consumer Affairs, Industry, Canada.

Gordon J. Lenjosek, Business Income Tax Division, Department of Finance, Canada.

Donald G. McFetridge, Department of Economics, Carleton University.

Jeffrey G. MacIntosh, Faculty of Law, University of Toronto.

Jack Mintz, Faculty of Management, University of Toronto.

Paul A. Toriel, formerly, Senior Market Manager, SMEs, Business Banking, Head Office, Royal Bank of Canada, Toronto.

Michael J. Trebilcock, Faculty of Law, University of Toronto.

David A. Wolfe, Department of Political Science, University of Toronto.

Larry Wynant, Richard Ivey School of Business, University of Western Ontario.

PREFACE

The conception of this book arose out of a May 1995 workshop held at the Faculty of Law, University of Toronto, sponsored by the Faculty of Law (Professor Martin Friedland, then Acting Dean); Faculty of Law, Law and Economics Programme (Professor Michael Trebilcock, Director); the Toronto law firms of Borden & Elliot and Stikeman, Elliot; and the Canadian Bar Association-National Competition Law Section. The idea for a workshop on 'Competitive Industrial Development: The Role of Cooperation in the Technology Sector' evolved after Robert Howse (Professor, Faculty of Law, University of Toronto) suggested that since Richard Braudo was going to spend the summer of 1993 in Dallas he should research the history and current status of SEMATECH, Inc. (the consortium of US semiconductor manufacturers) in nearby Austin, Texas.

While surprisingly little had been published concerning SEMATECH, Inc., thanks to Aaron Blumenfeld (Member, Law Society of Upper Canada and Litigation Associate, Borden & Elliot, Toronto), an introduction was obtained to Ari Reubin (Senior Manager, Personal Productivity Division, Texas Instruments, Inc., Dallas, Texas; formerly, Manager, Technology Transfer, SEMATECH, Inc.). Through interviews of SEMATECH personnel, the provision of in-house written material and a tour of SEMATECH's restricted access facilities, Ari Reubin facilitated a unique information gathering opportunity. With direction from Professor Robert Howse, this provided the starting point for Richard Braudo to write a research paper as a third year course, which became the springboard for the workshop and the basis for his contribution to the book.

Transforming the idea of the workshop into its actually taking place required organization well beyond the capacity of a Student-at-Law. Jeffrey MacIntosh (Professor, Faculty of Law, University of Toronto) expressed an interest in and made commitments to both organizing and contributing a paper to the workshop. The success of the workshop, including the availability of the papers and commentaries which are the foundation of this volume, are a result of the efforts of co-editor Jeffrey MacIntosh.

PREFACE

[illegible]

INTRODUCTION

As the global economy approaches the end of the twentieth century, transnational corporations (TNCs), small- to mid-sized enterprises (SMEs) and governments have emerged as the principal players on the stage of industrial development. In a dynamic knowledge- and technology-driven global marketplace, characterized by the increasing democratization of national governments, the leading suppliers and producers of services and goods cannot afford to lose sight of their paying audience: individual citizens and collective consumers constituting domestic, international and global consumer demand. This volume addresses salient issues arising out of R&D- and innovation-driven economic development during the past 25 years, and their implications for regional, national and global industrial growth (economic competition) in the mature Global Age of Information of the twenty-first century.

Part I provides an analysis of federal and provincial (Ontario) industrial policies. In Chapter 1, Robert Howse presents a critical analytical framework for evaluating recent federal technology-oriented industrial policy, focusing specifically on its cooperative content (as between the principal players). Similarly, focusing on the role of cooperative industrial policy, but with the more detailed approach of case studies, David Wolfe develops a constructive analysis of recent federally and provincially driven policy in Chapter 2.

A key lesson to be learned from their analyses is that with respect to both national and regional (federal and provincial) industrial policy, from a law and economics perspective, the design of such policy has been neoclassically driven. That is, the underlying rationale and purpose of such policies has been the internalization of (perceived) macroeconomic externalities (in respect of R&D-based innovation) and the subsidization of (perceived) fundamental (resource-based) industries to be achieved through the financing of firm-, regional-, and/or industry-specific enterprise development.

Federally (nationally), three conclusions are to be drawn: (1) industrial policy to date has been fundamentally misguided, in effect contributing little to economic growth but much to the growth of federal and provincial deficits; (2) properly understood, Canada has performed and continues to perform well in terms of global competitiveness; (3) to the extent that cooperative industrial policy initiatives are appropriate, they are best designed so as to be contextually oriented towards technology-based economic growth.

In the Canadian context this means the development of comparative advantages in the application of advanced technology in core resources (forestry, minerals, oil and gas), basic manufacturing (automotive and telecommunications equipment) and leading services (finance, information processing and telecommunications) for export, as well as domestic use or consumption. Provincially, conclusions to be drawn are fundamentally the same. The specifics of provincial industrial policy, however, must be tailored to the regional context in question. In the case of Ontario, for example, this would mean particular emphasis on basic manufacturing and services.

In the final analysis, based on theoretical and empirical analysis, Howse and Wolfe can be seen to conclude that it is time to develop and utilize a post-modern neoclassical economic conception of global economic reality. Economics has never presented itself as an exact science. Nevertheless, like any science, it is inherently defined by disciplined theoretical and empirical evolution. A fundamental concept in neoclassical economics is that of competition. In the Global Age of Information, economists and public policy-makers may be required to seek understanding of the post-modern global economy through the lens of a post-modern neoclassical conception of competition. Parts II and III provide theoretical and empirical analyses of the roles played by neoclassically based public and private sector financing and their implications for the planning and development of industrial policy. In Part IV the discussion turns towards a synthesis of the conclusions drawn from Parts I–III in terms of methodological, theoretical and empirical analysis. The result is an empirically based post-modern neoclassical conception of competition as cooperation in which significant, if not determinative, neoclassically perceived externalities are seen as internal parameters in the calculus of competition in a maturing Global Age of Information.

In Part II theoretical and empirical analyses of the impact of past Canadian and recent Australian federal tax policies provide evidence of the misguided nature of Canadian industrial policy as well as a model for the design and evaluation of more effective technology development tax policy. In Chapter 4 Paul Halpern and Jack Mintz suggest that to the extent interventionist policy is warranted, modern neoclassical theory dictates that tax incentives be provided to those areas of technology development in which Canadian enterprises have established or seek to establish comparative advantage(s). They point out, however, that Canadian tax policy has been designed, for the most part, to provide favourable treatment (incentives) to Canadian enterprises operating in traditional resource and manufacturing industries. Instead of promoting the development of comparative advantages, by orienting tax incentives towards industrial sectors in which comparative advantages have eroded over the last 25 years (due principally to the relatively high cost of labour), the end result is the allocation of scarce resources into less competitive, stagnant or negative growth industries at the expense of reduced investment in more competitive opportunities with positive growth potential.

In addition to such perverse sectoral bias, marginal effective tax rates (METRs) are significantly lower for SMEs versus TNCs. For example, in the communications sector METRs of SMEs are only 11 per cent of those of their TNC counterparts. Halpern and Mintz conclude that not only is the lowering of METRs ineffective in protecting against changes in the global composition of comparative advantage, but the higher METRs on more globally competitive enterprises with higher expected growth rates (in absolute terms microeconomically, and marginally in terms of contributions to overall GNP) may result as well in lost economic opportunities.

The underlying question raised by the discussion of Halpern and Mintz is the extent to which, if at all, tax incentives can be employed to effectively and positively impact technology development. Gordon Lenjosek considers this question in Chapter 5, a case study of the 1993 Australian Bureau of Industry Economics publication evaluating the effectiveness of a 150 per cent of eligible R&D expenditures tax deduction in meeting the policy goals of increasing firm investment in R&D, making firms more innovative and globally competitive; and showing whether such assistance has positively impacted economic welfare in Australia.

While the specific tax policy and its evaluation are interesting, more important for this discussion are lessons to be learned regarding appropriate methodology and effective statistical modelling applicable to the analysis of the design and implementation of tax policy related to industrial development in the Global Age of Information. In particular, Lenjosek derives the following lessons from his analysis:

1 Information obtained from literature surveys and case studies is useful.
2 Methodology for creation of statistical models used to evaluate the effectiveness of given tax policies must be carefully selected to provide for unbiased, accurate and meaningful statistical information.
3 Since the purposes of such modelling is to evaluate the effect of tax policy on industry-specific and industry-wide intensity of investment in R&D and their degree of innovativeness, and to determine whether any significant welfare spillover effects result, causality is of critical importance. In creating econometric models, the choice of independent and dependent variables must take into account theoretical, empirical

and statistical factors so as to provide the basis for concluding that significant statistical correlations are a result of underlying causal relationships.

In addition to tax policy, an alternative to internally financed R&D (typically only available to large, well-established firms) is bank financing of knowledge- and technology-based economic development. In Part III the issue of external financing for SMEs, leading players in technology development, is addressed. In Chapter 6 Jeffrey MacIntosh addresses theoretical issues with respect to the availability of bank financing to R&D-based SMEs. A principal reason for constraint expected in such lending is the fact that high technology SMEs characteristically lack traditional bankable assets (in particular, tangible assets and accounts receivable). Thus, R&D-based SMEs lack the traditional collateral banks require for lending. Nevertheless, to the extent that valuation advisers are beginning to include intellectual property and human capital as fundamental assets, banks may be expected to include both as components of collateral as the two elements become increasingly important in the Global Age of Information. Consequently, while bank lending to R&D-based SMEs continues to be constrained, it can be expected to become less constrained in the future.

Beyond increased risk associated with lack of collateral, MacIntosh argues that banks can be expected to perceive R&D-based SMEs to be subject to high systematic risk due to increased overall business risk associated with the R&D, product development and market research orientation of SMEs. These systematic risk factors typically engender high fixed to variable cost ratios resulting in R&D-based SMEs being subject to increased vulnerability to macroeconomic cycles. While this suggests increased vulnerability to recessions (negatively impacting financial performance), it would also suggest increased 'vulnerability' to upward swings in economic growth following recessions (positively impacting financial performance). Over the long term, such systematic risk should be discounted (assuming firms survive recessionary periods). Thus, the increased systematic risk that banks perceive R&D-based SMEs to be subject to may also be seen as arising from the failure to take into account the positive implications of associated risk factors, perhaps a consequence of the utilization of short-term horizons in loan portfolio management.

Finally, MacIntosh points out that a given SME can be expected to be perceived by banks as subject to high unsystematic risk stemming from (1) its focus on developing new products with an unproven market and (2) the inherent difficulty faced by banks in evaluating the quality of SME management (human capital). Given their lack of traditional collateral and perceived high systematic and unsystematic risk, MacIntosh concludes that bank lending to R&D-based SMEs can be expected to be seriously constrained. Acknowledging this conclusion, Paul Toriel, in Chapter 7, discusses the empirical significance of such constraints and ways in which they might be overcome.

Toriel points out that since 1981, technology-based SMEs have accounted for the majority of new jobs created in Canada and have played a leading role in Canadian economic growth. The Global Age of Information is increasingly characterized by knowledge intensity (even in resource and traditional manufacturing industries) and increased services sector domination (growing from 56 percent to 76 percent of Canadian GDP between 1971 and 1994), with competitive advantage in niche marketing (globally) a key to economic viability. In addition to facing constrained availability of bank financing due to factors identified by MacIntosh, in striving to capture niche markets globally such SMEs are more oriented towards international trade, the complexities of which are little understood by lenders and investors alike. In short, Toriel concludes that knowledge-based SMEs are faced with an increased need for external financing and a decreased likelihood of obtaining such financing.

In the midst of so difficult a financial environment, creative solutions are required. Any such solutions must redress informational asymmetries in the form of (1) failure to account for intellectual property and human capital in the calculation of bankable assets (collateral), (2) discounting of the positive impact of being more vulnerable to macroeconomic cycles, and (3) lack of understanding of international trade and global markets to which knowledge-based SMEs are oriented. Toriel's discussion of the Nat West Bank (one of the largest full-service English banks) and US Silicon Valley Bank (a small northern California commercial bank) models for financing SMEs is instructive.

Notwithstanding the fact that, due to the disparate natures of a large national versus a small private bank, the models are markedly different structurally, they share common elements of cooperation between business, lender/investor, government, and university networks. Such cooperation accounts for the success to date of both models. In this context, the cooperation is provided to assist a given SME, but not between business competitors and the other three networks. Toriel concludes that a new partnership is needed between business, academia, industry groups, financial institutions and governments: development of a 'positive conspiracy' to facilitate the financing of Canada's knowledge-based SMEs, the financing which is essential to their growth in the ever-changing economy of the Global Age of Information.

If cooperation between financial, governmental and academic networks on behalf of individual SMEs is the key facilitator of their economic viability in the Global Age of Information; and, knowledge-based SMEs play a principal role in job creation and a significant role in economic growth; then, a logical extension of the analysis is to consider the role of cooperation not only between a given firm and the three networks but also between networks of firms (traditionally perceived and typically legally required to compete with each other to the exclusion of cooperating with one another) and the other three networks (so to speak, 'systematic' as well as 'unsystematic' cooperation). Such consideration is the theme of Part IV.

In Chapter 9 Donald McFetridge argues that government accommodation of cooperative innovation between erstwhile competitors for global market share is increasingly important relative to government (financial) encouragement via tax incentive, lending and grant programs. He gives three reasons for the need to change industrial policy:

1 recognition by policy-makers of the ineffectiveness of government-led financial encouragement, particularly as a result of tax-subsidy approaches being discredited by guru (Porter) and scholar (Nelson) alike, who emphasize institutional linkages as a key to success;
2 the scope for domestic subsidy programs increasingly restricted by international trade agreements, with prevailing subsidies vulnerable to countervail; and,
3 government fiscal constraints restricting the ability to provide cash grants, below market interest loans and tax incentive write-offs.

Given that competition (anti-trust) and regulatory policy form the structural foundation of government accommodation, domestic and international legal frameworks, specifically competition and intellectual property regimes, need to be consistent with the accommodation of innovation.

Because, (1) cooperative R&D and innovation efforts (in the form of strategic alliances, joint venture agreements, etc.) can necessitate the exclusion of future competitors and may be perceived to reduce future competition, such explicit or implicit lessening of competition, and (2) such cooperation is essential to effective R&D and innovation (knowledge- and technology-based enterprise growth), McFetridge concludes that any such cooperative effort should be evaluated in terms of whether the resulting innovations would have likely occurred or would have occurred in as timely a fashion in the absence of an accommodating competition and regulatory policy. Unfortunately, to date, in both the US and Canada (according to US Department of Justice-Federal Trade Commission Antitrust Guidelines for Licensing of Intellectual Property and the Draft Information Bulletin on Strategic Alliances under the Canadian Competition Act, respectively) such cooperative efforts continue to be viewed, a priori, as anti-competitive, civilly. (In Canada, criminal liability is also a possibility.)

McFetridge's discussion and conclusion suggest a fundamental change in the orientation of domestic competition and regulatory policy: a shift in focus from national/international markets to global markets (i.e. viewing domestic competition as a function of global competition). In Chapter 10 Richard Braudo provides a historical, empirical and theoretical analysis of the compelling need for changes in industrial policy-making and, more fundamentally, the prevailing conception of competition. Reaffirming conclusions reached in Parts I–III, especially the fact that microeconomic and macroeconomic growth in the Global Age of Information is increasingly driven by R&D and innovation (knowledge and technology), and, based on empirical research (literature search and original research in respect of SEMATECH, Inc. in Austin, Texas), microeconomic and macroeconomic theory and analysis, and modelling methodology consistent with fundamental econometric principles, Braudo concludes:

1 the cart may have been put before the horse in reinventing government prior to reinventing competition; and,
2 most likely, if not certainly, as we move into the twenty-first century, the Global Age of Information calls for the conceptualization of competition as a much more complex and dynamic, multidimensional phenomenon than simply the driving force of so-called 'free (unregulated) markets', *laissez-faire* or 'Darwinian' capitalism.

Accordingly, rather than the fixed conception of competition underlying free market (modern neoclassical) economic thinking, the New Competition is conceived as being four dimensional:

1 global, regional and domestic markets, usually but not necessarily interrelated;
2 markets for capital (liquid capital in the form of cash and securities, fixed material goods and resources, intellectual property, and human capital in the form of labour and management), goods and services, and transnational corporations (TNCs);
3 global, regional, domestic and local demand (consumer, producer, quasi-public and public goods and services);
4 time, relative to which elements of each dimension or the dimensions themselves may change, i.e. the key elements of dimensions 1–3 specified above are acknowledged to require periodic review (periodicity based on the rate of change during a given policy period) as to their appropriateness over time.

Replacing the modern neoclassical competitive imperative of winning free (ideally unregulated) market share (i.e. being Number One) is a post-modern neoclassical global competitive imperative of becoming and remaining a viable cooperatively competitive participant in the global market.

To replace the neoclassical posture of a virtually bygone era, the Global Age of Information calls for the forging of a new partnership of co-operation between TNCs (macroeconomically speaking, the principal economic players on the global stage of knowledge- and technology-driven economic growth) and government(s); cooperative efforts through which economic goals based on a newly understood global competitiveness are achievable. Just as significantly, academics, policy-makers and their advisers, leaders in business and finance, as well as consumers, must incorporate in their decision-making calculus the fact that the elements of dimensionality are not fixed over time. Rather, elements of the New Competition are potentially as fluid as the global movement of various forms of capital, products and services, not to mention changing forms of government.

The case study of SEMATECH provides empirical grounding for the theoretical analysis and informal model developed. It should be emphasized that the SEMATECH example does not provide confirmation of the New Competition model, but the theoretical analysis and informal model developed are consistent with the facts and experience of SEMATECH to date. Therefore, lessons derived from SEMATECH, while not conclusive, are instructive. In addition to cooperation between (1) erstwhile competitors for global market share and (2) government, academia and those same commercial global competitors; the very existence (the result of enabling Congressional federal legislation) and operating structure of SEMATECH, Inc. are the embodiment of the sort of accommodating competition and regulatory policy suggested by McFetridge.

One lesson of particular interest, given the key role played by R&D and innovation in economic growth, is the SEMATECH approach to intellectual property. Modern neoclassical economic analysis perceives the need for a patent law regime in respect of R&D and innovation because private and social needs (and incentives) are seen to be divergent, leading to public good attribution and the attendant free-rider problem. However, the SEMATECH approach to intellectual property and its evolving treatment of patent rights for innovations developed, moving away from exclusive rights and towards 'common' market rights, suggests that the patent law component of competition and regulatory policy requires careful reconsideration. At least where cooperative R&D and technology transfer are concerned, the SEMATECH case indicates that the global competitive imperative renders restrictions on licensing on the one hand, and patent breadth and length on the other, inappropriate. This is especially apparent where product cycles are short and first mover advantages great, as is typically the case in high-technology industries. Contrary to much current thinking, prevailing intellectual property regimes in Canada and the US may suffice, though the statutory definition of patent length and limitations on licensing may be unnecessary and at times counterproductive to innovation and growth.

Braudo concludes, therefore, that competition and regulatory policy which adopt a rule of reason approach, in which competitiveness is presumed in favour of national or regional cooperative R&D ventures, appears to be more appropriate in the context of knowledge- and technology-based economic growth in the Global Age of Information. Moving towards the twenty-first century, the Global Age of Information calls for an understanding of competition as a more complex and dynamic, multidimensional phenomenon, taking on different forms, depending on time, place and circumstances. Taking all four dimensions into account, SEMATECH teaches that where home-based TNC and domestic interests coincide, cooperation may significantly lessen domestic competition at certain stages in the continuum of bringing products or services to market. While spillover effects and market failure corrections may be added benefits of such cooperation, they are not the *raison d'être*. In respect of the domestic market *per se*, such cooperation may be essential to the global competitiveness of domestic industries (TNCs and SMEs alike). In addition to overall competitiveness, SEMATECH suggests that such cooperation is a fundamental factor due to the reality of competition between governments and nations/regions for the location of TNCs. The New Competition necessitates the addition of government as an element of national/regional competitiveness, meaning government policy itself is an endogenous factor in the global competition platform. Unfortunately, in this regard, Canadian competition policy as embodied in the Competition Act (amended 1985, 1990), in addition to problems pointed out by McFetridge, incorporates a merger regime diametrically opposed to the type of regime which would accommodate the kind of effective cooperative industrial policy suggested by SEMATECH. It is derived from the conventional, modern neoclassical conceptualization of competition. While US courts have acknowledged the global competitiveness imperative (as in *General Dynamics*), in addition to problems with the US regime pointed out by McFetridge, where US policy is concerned the onus remains on the merging parties.

Competition and regulatory policy which recognize the New Competition and its attendant need for cooperative industrial policy should incorporate a rule of reason approach in which the onus would be on plaintiffs to prove the anti-global competitiveness of the cooperative effort in question, taking into account relevant elements of each component of the New Competition's dimensionality. Once competition is understood to include cooperation, modern neoclassically based, government-dominated and domestically myopic competition and regulatory regimes and their accompanying industrial policies of attempting to choose or provide incentives to potential winners may be replaced with cooperative, globally oriented networking of members of business, finance, government and academic sectors to facilitate sustained economic viability (competitiveness) in the Global Age of Information.

Given the methodological, theoretical and empirical lessons of the discussion in this volume, substantive law and economics insights are provided for the consideration of academics, policy-makers and their advisers, regulators and judges, as well as business leaders, entrepreneurs and consumers. Nevertheless, the more significant lesson is the need for additional research. Empirical work utilizing more formal modelling and statistical (econometric) analysis is needed to test the validity of the New Competition conceptualization for the Global Age of Information. Such research will confirm or disconfirm and provide the basis for adjustment and formalization of a model of competition more useful to industrial policy-making in the twenty-first century.

Part I

COMPETITION AS COOPERATION

Policy perspective

1

HIGH-TECHNOLOGY INDUSTRIAL POLICY IN CANADA

The future of an illusion or an illusion of the future

Robert Howse

INTRODUCTION[1]

This chapter is intended as a critical analysis of a claim frequently made nowadays in discussions about the role of government in economic policy. Perhaps the clearest articulation of the claim that I have found in the literature is the following statement by David Crane, an economics journalist and prominent advocate of activist industrial policy:

> If technology, in the form of innovation, ideas and knowledge, is a key factor in economic growth then it follows that governments can do a great deal to facilitate growth. . . . In addition [to policies that support education, training and basic research] governments have a vital role to play in facilitating pre-competitive research and development with industry and in assisting the actual commercialization of new goods and services. The risks are so high in the development of new technologies today that companies often cannot undertake these efforts on their own; similarly the spillover effects of new technology can be significant so that companies making the expenditures on technological development may not capture all or a large part of the benefits. . . . The importance of risk sharing between industry and government is well understood in Japan, the United States and Europe but much less so in Canada.[2]

Like most analysts of industrial policy (IP) who are influenced by the neoclassical economic paradigm, I have generally expressed deep scepticism about this kind of claim.[3] However, I recently came across an article by one of the most thoughtful neoclassical economists of our time, Paul Krugman, in which he seemed to repudiate much of his earlier work on industrial policy, claiming that he and other neoclassical economists had significantly underestimated the significance of external economies, or externalities, and the potential, *at least in theory*, for industrial policy to internalize these externalities.[4] In particular, Krugman has been impressed by the evidence presented by Michael Porter, among others, about the importance of geographical clusters in economic growth or development. This evidence demonstrates the extent to which individual firms do not stand or fall by their own efforts but depend on the mutually reinforcing effects of each other's success.[5] Nevertheless, Krugman concludes his paper with the caveat that, while stating a theoretical justification for industrial policy may be relatively easy, the 'hard issues' are those of choosing the right instruments and preventing new industrial policies from being hobbled by the same mistakes that beset the old ones.

I wish to begin more or less at the point at which Krugman ends. Accepting Krugman's claim that state-of-the-art economic research supports the view that external economies are significant, what kinds of government policies are likely to be successful in internalizing them? Given that governments probably do not have the resources to internalize all external economies, and given that these external economies are relatively pervasive, how do we choose which to internalize? And, not of least importance, are the indicated policies likely to achieve their intended results in the Canadian context, given the historical experience of the political economy of industrial policy in this country?

EXTERNALITIES, PUBLIC GOODS AND FREE-RIDERS

Perhaps the most theoretically coherent rationale for government support of industries is that of positive externalities. Particularly in the high technology sector, it is argued, the innovative activity of one firm is likely to have significant spillover effects for other firms or other sectors. Advocates of IP argue that government action is necessary to internalize positive externalities and to ensure a socially optimal level of investment in firms that produce such externalities. They draw on an impressive and increasing literature on the positive linkages between firms and industries in successful regional or national economies.[6]

The IP literature often makes the error of identifying the existence of positive externalities as itself a demonstration of market failure. However, the more precise issue is whether significant externalities exist that cannot be internalized by a properly functioning market. When firms confer benefits on other firms, the value of these benefits will normally be internalized when the benefits can be proprietized and a market price charged to the firms that receive them. The market fails only where the firms that receive the benefits are able to free-ride.

In the case of innovation, or R&D, intellectual property rights are, of course, the principal legal instrument that addresses the problem of free-riding. However, not all the benefits that flow from research and development can be proprietized through intellectual property rights but only those that fall within legal definitions of what can be patented, copyrighted, and so on. While one response to the continued existence of free-riding could be to expand intellectual property protection, the current approach to such protection, with its winner-take-all characteristics, may lead to 'racing'; that is, two or more firms devoting resources to the developing of a single innovation in order to be first across the finish line. Moreover, as Itoh, Kiyono and Okuno-Fugiwara note, 'the patenting system involves an inherent contradiction in that while it raises *ex ante* R&D incentives by extending property rights, it also renders the diffusion of the developmental fruits, socially desirable in an *ex post* sense, difficult'.[7]

In addition, significant externalities may exist with respect to non-firm specific training and know-how that is conferred on workers. Firms may try to internalize as much as possible the benefits that flow from education and training through long-term employment contracts or covenants that restrict the capacity for workers to work for competing firms, but contract law rules themselves may significantly constrain such strategies.

The theme of this conference – cooperation and industrial policy – suggests that possible solutions to these problems may be found in cooperative arrangements among firms. For instance, if a particular R&D project promises to confer benefits on several firms but insufficient benefits on any one firm to justify that firm undertaking the project, the solution may be a cooperative arrangement of sharing of the project's costs among the firms which stand to benefit. Under some circumstances, the possibility of joint gains may justify not simply a joint venture but a merger.

Where the gains of cooperation, or the positive externalities at issue, are broadly spread across firms and sectors, it is rather easy to see why collective action problems might exist that would prevent firms themselves cooperating so as to share costs and benefits and internalize externalities. Indeed, neoclassical economists have long recognized the 'public goods' character of basic education and highly generic R&D as well as of some training.[8] However, where significant linkages exist among a relatively small number of firms or sectors, it is less clear why activist government policies would be needed to achieve the desired cooperative outcome. It is often argued that some forms of inter-firm cooperation may need exemptions from anti-trust or competition laws, as was the case with the American SEMATECH venture to be discussed by another panel at this conference. However, beyond removing this kind of legal obstacle to cooperation, it is unclear why government would need to be active, for example, by providing subsidies or research facilities.

One justification sometimes offered for a greater government role is that of information failures. Individual firms are prone to make investment decisions in ignorance of how these decisions may affect other firms. Since developing proprietary research makes secrecy important, firms may not disclose to each other enough information about their planned or existing activities to make clear the possible gains from cooperation. It is arguable that a government agency staffed with highly qualified analysts, scientists and engineers may be able to discern linkages and opportunities for cooperation to which individual firms are blind. The agency would have access to a vast amount of information about R&D and new technology being developed throughout the economy. This kind of role has often been attributed to MITI, the Japanese Ministry of International Trade and Industry.[9] According to Porter,

> MITI conducts or commissions countless study groups, industry committees, and reports concerned with new technologies, trends in international competition, and future issues. These are conducted with the input of the best Japanese experts, academics, high-level industry representatives, and government officials. The reports are broadly disseminated and publicized and are widely covered in the press. The major function of such studies is to awaken firms to emerging trends and problems.[10]

Another bar to cooperation between firms that captures positive externalities is the transaction costs of cooperative activity. Even when a cooperative activity yields joint gains, each firm may be concerned that the other will reap the larger gains from the results of the cooperation. If the consequence is that the other firm will gain a relative competitive advantage it did not have before cooperation, then a concern about 'relative gains' may dissuade some firms from entering into cooperative arrangements that promise to yield 'absolute gains' to all the participants.

Aside from concerns about 'relative gains', concerns about cheating may also deter firms from entering cooperative arrangements. For instance, firms may worry that they will end up giving competitors access to trade secrets or proprietary research. The most advanced firms in an industry or in a technology will be concerned that other firms will free-ride within the arrangement and so will be less likely to provide their best scientists and engineers. Indeed, the evidence of attempts at inter-firm collaboration in high-tech R&D in Japan suggests that distrust between firms has been almost universally present initially, and that only after sustained efforts has effective collaboration been possible.[11] Government involvement in the formation of such cooperative arrangements could be one response to the transactions costs posed by relative gains concerns and by the possibility of cheating. Partial government funding of research could be viewed as compensation to firms in the syndicate for the risk that the ultimate research result will end up primarily benefiting the other firms. Moreover, government may be able to intervene more directly, for example, by punishing or by threatening to punish firms that cheat, by withholding other government benefits or threatening to exclude them from future cooperative endeavours. It is quite conceivable that this latter role of government as arbitrator and enforcer could substantially reduce the transaction costs of cooperation.

Third, it is often argued that capital market imperfections may lead to underinvestment in high-risk, high-cost R&D-based ventures, even if cooperation succeeds in the capturing of significant external economies. Such ventures may simply be too risky given the risk-averseness of most investors. It is also claimed that managers of pension funds or mutual funds are under pressure to maximize short-term returns and will find investment in ventures that are based on high short-term costs, with returns taking many years to be realized, unattractive. Finally, it is claimed that the sheer volume of capital needed to undertake some high-tech ventures is simply not available on the market on any terms.

As an investor, government supposedly is not under the same constraints as capital markets. Because it can raise a large quantity of capital from tax revenues, it can supposedly do an end run around the risk-averseness of private investors. However, I am sceptical of characterizing the unavailability, or scarcity, of capital for very high-risk ventures as a market failure or imperfection. To say that investors do not want to take the risks is simply to say that, based on investors' *revealed preferences about risk*, the costs of capital are prohibitive and the venture is not efficient. (Again, if the claim is that social efficiencies are not captured in the projected return to potential investors we are back to the externalities-based arguments.)

An important qualification of this statement would be if one could show that investors' perceptions of risks or risk/return trade-offs are affected by heuristic biases. Some evidence in recent psychology literature, for example, suggests that people systematically overestimate risks where they are required to make choices about risk under conditions of considerable uncertainty.[12] Another qualification might be that, because of the large amounts of capital at its disposal, government can diversify risks in ways that are difficult or impossible for private investors. However, given the possibility of diversification through mutual funds or portfolio techniques of investment this claim appears implausible.

Finally, in some situations government may have lower agency costs of capital than private investors. Depending on the culture of government–industry relations in a particular society, firms and their managers may be prepared to disclose more information to governments than to private investors. The Japanese government is said to have enormous leverage that it can exercise over the companies it invests in. It is able to reward and punish firms and their managers in ways that would be all but impossible for any private provider of capital. In high-tech ventures, where considerable amounts of capital must be locked in for long periods of time before returns are realized, these informal kinds of 'voice' may be extremely important in controlling agency costs. Needless to say, however, exercising this 'voice' effectively depends on the government having the institutional capacity to monitor closely the ventures it invests in. Students of Japanese industrial policy argue that, in MITI, the Japanese government has assembled just such an institutional capacity. And, indeed, empirical evidence suggests that, in the case of high-risk ventures in which the Japanese government has invested considerable resources, MITI tends to micro-manage, or at least micro-monitor the venture, aggressively demanding changes in strategy or personnel if results are not to its satisfaction.[13]

WHAT'S SO SPECIAL ABOUT HIGH TECH?

Advocates of targeted high-tech industrial policies claim that the current range of policies and government resources devoted to supporting technological innovation is inadequate. Showing that externalities exist, as Krugman suggests, may not be difficult. However, showing that existing market institutions and government policies do not provide for sufficient internalization of these externalities is much more difficult. Besides, IP advocates are interested in convincing policy-makers, and the internalization of external economies as an end in itself is of little interest to most Canadians, although it undoubtedly excites professional economists. This leads most industrial policy advocates to sidestep the issue by adopting instead a comparative benchmark of how 'we' as a country are doing in comparison to others. Usually the claim is that we are falling behind some other country, whether Japan, Germany, or the United States, both in our policy responses to the high-tech challenge and in our overall economic performance. The citation of both claims together often substitutes for any causal analysis of how the second is the result of the first.

The crudest way of stating the claim that existing policies and institutions are inadequate is to compare Canadian spending on research and development or the technology-intensity of our exports to the 'performance' of other countries. This comparison almost invariably provides a grim picture of the Canadian situation. For instance, of the total export market shares of seven OECD countries in 1986, Canada made up only 3.5 per cent in the 'High Intensity R&D' category (as a point of comparison, the US share was 22 per cent). Canadian exports accounted for 5.5 per cent of the 'Low Intensity R&D' category with American exports accounting for 8.5 per cent.[14] Using export receipts from the sale of technology itself, the Canadian performance is even more lacklustre – among the same group of OECD countries, Canada accounted for only 1.9 per cent of the total.[15] Moreover, there is also a dramatic trend, beginning in the early 1970s, showing a decline in the use of domestic Canadian technology in Canadian manufacturing, with a drop of 27 per cent noted for the period from 1972 to 1984 alone.[16] Total Canadian spending with respect to research and development (with private and government spending combined) has lagged behind that of several other industrialized countries, including the United States.[17]

To those Canadians dazzled by the ability to produce gadgets and contraptions, these figures lead to a sense of wounded national pride. However, they must be examined side by side with different figures that suggest Canada's rate of economic growth and our export performance over much the same period have been relatively positive, including when compared to other countries whose industry is characterized by a much higher level of production of technology-intensive products. For example, during the 1975–1987 period, a time of greatly increased competition for export markets among industrialized countries and between NICs and industrialized countries, Canada was the only country besides Japan in the group of seven OECD countries to actually increase its share of world exports.[18] Measured against members of the UN, a much larger group of countries, Canada's share of total exports remained between 4 per cent and 5 per cent from 1950 to 1989, a period in which the US share declined from 18 per cent to 14 per cent and Japan's share increased from 2 per cent to 8 per cent.[19] In its overall levels of economic growth, during the 1980s and 1990s Canada has lagged behind other OECD countries at some times and outpaced them at others. If we use an even more general indicator of Canada's economic performance, one which takes into account factors such as the quality of life that the Canadian economy is able to sustain, our comparative achievements appear even more impressive. Thus, in the United Nations Development Program *Human Development Report*, Canada has consistently ranked first or second among all countries in recent years.

That Canada's comparative performance has been extremely good, however measured, while our dependence on domestic technology has declined and the technology-intensity of our exports remains low, justifies some initial scepticism as to the case for targeted high-tech industrial policies. This is not to say that, in absolute terms, Canada's economic performance could not be improved. In particular, persistently high levels of unemployment in depressed regions of the country are a genuine problem. Yet it is in this last area that government policies have been most intensive but have largely failed in solving the problem. Another problem, also experienced by countries with much higher levels of government investment in high-tech industry than Canada, is chronic unemployment among unskilled or semi-skilled workers, especially young people.[20] Here, also, it is far from clear that a targeted high-tech industrial policy would result in the kind of economic growth that would put unskilled workers to work.[21] Indeed, better internalization of technological externalities might actually accelerate the processes of technological change that result in capital's replacement of labour, making the problem of unemployment among the unskilled even worse.

More fundamentally, the fact that Canada remains a successful exporter while its domestic production of high technology and technology-intensive products remains relatively low suggests the possibility that many external economies or positive spillovers from such production flow *across national borders*. Niosi remarks, for instance, that large Canadian companies have become adept at buying technology, plugging into R&D activities abroad, and adapting foreign-made technologies to their industrial requirements. He suggests that Canadian multinationals are 'among the world's fastest "followers"'.[22]

To the extent that other countries maintain substantial state support for research and development and for industries that produce technological innovations from basic research, Canadian skill at what Niosi calls 'following' will allow us to benefit substantially from free-riding on these other countries' policies.[23] Therefore, when David Crane and other advocates of high-tech IP assert that we are falling further behind the Japanese or the Americans in these areas, they invite the possible response that our policies in fact enhance our opportunities for free-riding, and that, under such circumstances, augmenting our own government support for high-tech would be an entirely perverse policy shift. In sum, the more that others spend on these things, the more we should reduce our own spending.

If, indeed, external economies and spillovers flow across national borders, it also follows that our own efforts to internalize these economies through national policies may well be largely futile. As Cynthia Beltz notes,

> there is no reason to presume that spillovers will honour politically defined boundaries. Advances in communication and transportation are tying trading partners closer together; distance is becoming less important; and, with research and development occurring in a variety of nations, industries are becoming more geographically dispersed, further facilitating the spread of technological spillovers.[24]

Producing more government-funded non-proprietary research in Canada, there is little guarantee that the positive spillovers from such research will flow largely to Canadian as opposed to foreign firms (indeed, even where such spillovers flowed to Canadian firms they might lead to increased production by such firms offshore with little additional employment created). And if others are able to free-ride on our own government-supported research and technology production, they are likely to behave rationally and reduce their expenditures in these areas, eventually decreasing our own opportunities for free-riding.

An obvious objection to this argument is that it implies that governments in countries such as the United States, Germany, and Japan are irrational in devoting considerable public resources to high-tech industrial policies when substantial possibilities for free-riding by other countries exist. Yet this is not necessarily the case. First of all, a superpower such as the United States is probably compelled to invest significant public resources into high-tech R&D and product development for reasons of national defence. Given the inevitability of such spending it probably makes sense to make a further investment in order to maximize the positive spillovers to the domestic civilian economy from R&D defence spendings. Second, the larger and more diversified the domestic economy, the more likely that at least a significant portion of the positive externalities created by government support will be captured within the economy. Industrial policies that make sense in the United States or Japan might make little sense in a smaller economy such as Canada's. Third, a significant element of 'path dependency' may be at work in the industrial policy choices of other countries – for instance, manufacturing industries in the US have traditionally been heavily dependent on domestic US innovation and R&D, and they may experience significant difficulties in maintaining competitiveness through adapting processes and technologies developed abroad.[25] Canadian industries, by contrast, having been long dependent upon foreign R&D, may have already done the 'learning' needed to take full advantage of the trends in globalization of R&D noted by Beltz.

FROM THEORY TO PRACTICE IN THE CANADIAN CONTEXT

Given the theoretical case for targeted high-tech industrial policy as well as the limits of that case, this section considers the likelihood that, *in practice*, government intervention in the Canadian context is likely to be successful in correcting or perfecting the market. I have already suggested in an earlier section one major qualification on the theoretical case for government intervention in the Canadian context; namely, that many externalities cross national boundaries making Canada, or Canadian firms, a net beneficiary of free-riding, given Canadian adeptness at appropriating foreign technology.

This section addresses a different set of concerns with targeted industrial policy. Advocates of the 'new' high-tech industrial policy are quick to distinguish their proposals from traditional Canadian industrial policy, which they usually admit has largely failed on its own terms to create sustainable growth or jobs. However, they rarely defend this claim in any depth, although it is possible that the same political and institutional factors that may have led to the failure of earlier industrial policies are somehow irrelevant to targeted high-tech industrial policies.

First of all, it is simply not true that in the past Canadian industrial policy has been about protecting losers rather than picking winners. Some of the most spectacular failures of Canadian industrial policy have been attempts to pick winners – the Briklin sports car, Consolidated Computer, and Novotel (cellular telephones) are three examples.[26] A 1989 federal government publication lists 53 pages of federal (and some provincial) programs that provide support for technology development.[27] Contrary to the suggestion of IP advocates that (in the words of one such advocate), 'Canadian governments at all levels have been rather inactive', the 1989 publication describes dozens of programs and incentives that exactly fit into the supposed paradigm of 'new' industrial policy, including incentives for innovation, for diffusion of technology, and for private/public partnerships. Although the total percentage of Canadian GNP spent on R&D is relatively small, the proportion of this total attributable to government spending is one of the highest in the OECD.[28] Is the supposed inadequacy of these programs attributable simply to insufficient funding or to mistakes in design that can be fixed easily based on superior knowledge or 'learning'? This is, in essence, the claim of high-tech IP advocates. But it is a claim largely unsubstantiated by detailed analysis of past or existing Canadian policies. Curiously, although many of these IP supporters present themselves as Canadian nationalists, they are too busy looking into foreign policies to spend much effort carefully studying those that Canadians have themselves come up with. Without adopting the extreme public choice perspective that all government policies are subject to capture by powerful interests or are ultimately the product of marginal-vote maximizing conduct by politicians,[29] it is nevertheless possible to question whether the past record of politicization of targeted industrial policy decisions in Canada does not bode ill for the 'new' industrial policy as well. To return to the paper by Krugman discussed at the beginning of this chapter, in stating the economic case for industrial policy Krugman adds the crucial qualification that such a policy will only vindicate itself if the 'political economy' of state aid can be appropriately managed.

Meric Gertler, a strong advocate of high-tech IP, observes with respect to the targeting of industrial policies in the past in Canada: 'The Prime Minister's Office orchestrates the regional location of major federal contracts for defence-related aerospace and shipbuilding manufacturing and servicing to serve largely political purposes.'[30] It is unclear why Gertler believes that a more activist or expensive policy would not be subject to the same politicization. Again, I do not wish to take the position that politicization or capture is inevitable; in some cultural or institutional contexts it seems to have been possible to build an industrial policy apparatus relatively autonomous from pork-barrel politics.[31] However, advocates of high-tech IP, while acknowledging that such policies in other countries are dependent on a different political culture and institutional setting, tend simply to wish away the problem of capture or politicization that occurs in our own democratic system. At a minimum it is necessary to suggest politically feasible (at least in the long run) changes to government decision-making processes and institutions that would increase the likelihood that politicization and capture will not plague the new industrial policy as they have the old.[32] In any case, it would probably be folly to undertake a new high-tech IP, or, more precisely, throw more resources at it, without having achieved important institutional innovations in government–industry relations. Certainly, it is hard to identify any intrinsic characteristic of high-tech industrial policies that makes them less susceptible to pork-barrelling than earlier industrial policies were.[33]

One particular dimension of the 'new' industrial policy may put its success even more at odds with Canadian political culture than the old industrial policy. In focusing on external economies, a major rationale of the new industrial policy is to encourage or strengthen regional or local linkages between industries and firms, Michael Porter's 'clusters'.[34] However, federalism values have always pushed Canadian industrial policies in the opposite direction – that of geographic diversification (as opposed to concentration) of industrial activity. Take any major federal defence procurement contract, for instance. Given the politics of federalism, the federal government feels compelled to parcel out various bits and pieces of the contracts to different firms in different parts of the country. If anything, this practice discourages the diffusion of technology and know-how throughout geographic linkages and clusters. Consider the political fallout if the federal government were to announce that it was going to channel all of its expenditure on technology into developing a 'Silicone Valley' on the outskirts of Toronto. While such a decision would be in keeping with the logic of linkages and external economies that drives the new industrial policy, it would be totally out of the question given the logic that drives the politics of Canadian federalism.

Even if the federal government had the nerve to undertake a genuine strategy of encouraging geographic linkages, the beggar-thy-neighbour competition for investment that characterizes the industrial policy of most Canadian provinces[35] could easily undermine these gains. Provinces would continue to offer large incentives to lure firms away from their 'clusters', threatening to destroy any external economies created by the 'new' federal industrial policy. Moreover, provincial policies tend to reproduce within a smaller geographic area, the same kind of diversification that has plagued federal policies – political pressure to spread public investment around the province is indifferent to the importance of maintaining and reinforcing local clusters. Indeed, inasmuch as one can say that making clusters even more effective in the production of external economies is the key to the new industrial policy, in the Canadian context the best first step would be to dismantle completely the old policies that impede or counter the natural formation of clusters or local concentrations through the operation of market forces.

If we examine the specific roles that government can play in addressing the existence of externalities, such as in reducing the transaction costs of collaboration among competing firms in an industry, they tend to assume that government can use its leverage to control or deter opportunistic behaviour that may impede cooperation or, in other contexts, result in high agency costs of capital. Yet Canadian industrial policy has traditionally been characterized by very weak monitoring or direction of government-funded projects. Canadian managers, while open to receiving government assistance, resist forms of assistance that involve direct bureaucratic involvement in or scrutiny of firm decision-making.[36] Nor have Canadian governments been successful in translating government aid into leverage on firm behaviour, which seems to be a pervasive element in Japanese industrial policy successes. Porter *et al.* note:

> Canadian companies have looked for government assistance in export promotion, investment in specialized infrastructure, government procurement, and other forms of support. They have, however, rarely cooperated with governments in areas that have important impacts on international competitiveness, such as R&D, training and education.[37]

Again, I am sceptical that a new policy, or more handouts to a different set of sectors, can break this pattern which seems deeply engrained in the Canadian culture of business–government relations.

Finally, it is far from certain that, even if the government could turn subsidies into leverage, it would be well placed to exercise this leverage. Attempts to consolidate expertise and information on industrial policy within a single bureaucratic agency have not, in Canada, been particularly successful.[38] Few of our best scientists, engineers and managers have been attracted to bureaucratic functions; nor has the hierarchy and seniority system in the bureaucracy lent itself to frequent or easy movement between the public and private sectors. Advocates of high-tech IP may respond that, with enough resources and clout being conferred on an industrial policy agency, it *would* attract 'the best and the brightest'. However, one of the major impediments to allowing such an agency considerable clout is the high degree of political control over discretionary spending required by the political economy of federalism in Canada, to which I have already referred.

As an antidote to the view that we can somehow will into existence a different institutional framework that would be congenial to high-tech IP, it is worth studying the literature on the industrial policies of countries, such as Japan, thought to have been successful in producing the needed institutional capacity within government. What the literature reveals is that, far from being the brainchild of some industrial policy genius, institutions like MITI developed through largely *ad hoc* responses within a particular social and cultural context to the challenge of reconstructing an economy devastated by war.[39] I am not endorsing cultural determinist theses that Japanese or European approaches cannot work in North America; indeed Japanese multinationals have been quite successful in importing certain management and production practices into North America. Rather I am suggesting that different countries' industrial policies often exhibit a kind of path dependency and that it is not so easy to jump on to someone else's path.

CONCLUSION: WHAT *SHOULD* GOVERNMENTS BE DOING TO LINK TECHNOLOGY AND ECONOMIC GROWTH?

The analysis up to now may appear to some as just another example of simplistic and fatalistic 'Chicago school' thinking – market failure is preferable to government failure, policies are inevitably captured, etc. In fact, I am a firm believer in strong, progressive government and also think that a range of desirable government policies can facilitate the exploitation of external economies and mutually beneficial linkages between companies and industries in the high-tech sector. For example, non-firm specific education and training have a number of characteristics that make them the most clearly indicated industrial policy in the Canadian context. First of all, because education and training create externalities that are widely dispersed throughout the economy, they are least able to be internalized through spontaneous collaboration between relatively small numbers of agents. Second, the externalities from non-firm specific education and training are likely to be more easily captured within the domestic economy; whereas skilled workers, scientists, etc., are undoubtedly mobile, they are less mobile than disembodied research and ideas. Third, given the argument in the first part of this chapter that Canada should continue to free-ride to a large extent on other countries' public investments in high-tech industrial policies, education and training are required for us to be effective free-riders, or 'followers'. Fourth, although cynics often protest that education and skills do not create jobs, in fact they do. They are major factors, according to Michael Porter himself, in the development of effective clusters that produce economic growth.[40] Fifth, non-firm specific education and training policies, while not immune from problems of capture and the politics of federalism, are markedly less so, since they need not entail large amounts of discretionary spending or the necessity of picking and choosing firms and regions that benefit. Moreover, some relatively straightforward means could be adopted to make such policies even less susceptible to capture and the politics of federalism; for example, instead of distributing education and training dollars to provinces or to institutions within particular provinces or regions, the federal government could provide vouchers that the recipient could use to obtain training or post-secondary education anywhere in Canada. Finally, apart from their value as an industrial policy, government-supported education and training may have important independent justifications in terms of equality of opportunity and distributive justice.[41]

A second set of desirable government policies are policies that preserve or enhance 'social capital'. In a recent study of Italian politics and society that has attracted considerable attention, the political scientist Robert Putnam has argued for the importance of 'social capital' as a basis for the success of a variety of collective human endeavours.[42] The main form of social capital consists of 'networks of civil engagement'. Putnam suggests: 'Networks facilitate flows of information about technological developments, about the creditworthiness of would-be entrepreneurs, about the reliability of individual workers, and so on.' Innovation depends on 'continual informal interaction in cafes and bars and in the street'.[43] He further notes: 'Like all public goods, social capital tends to be undervalued and undersupplied by private agents.'[44]

Much can be done to supplement or reorient existing policies to create or reinforce social capital. Among the major threats to social capital in our society is the lack of adequate child care. Exhausted and often overwhelmed by the dual responsibilities of work and children, most adults today have little chance for the 'continual informal interaction in cafes and bars and in the street' to which Putnam refers in the Italian context; child-care responsibilities also leave many parents with little time for skills upgrading or development that is not provided on the job. Child care represents perhaps the single most important policy challenge to preserving or enhancing social capital. As well, delivering social and adjustment policies through community-based groups and organizations tends to reinforce the social capital that exists in particular communities. Indeed, these policies may be more effective in their own right than those which are delivered through traditional bureaucratic structures.[45]

This last point reinforces the importance of evolving or reforming existing policies in response to the challenges of the new economy. Under-investment in high-technology or other sectors in which firms can generate significant positive spillovers may be just as much due to government failure as to market failure. Traditional industrial policies in Canada have often favoured primary (resource extraction and processing) industries.[46] This bias can be seen in tax incentives and approaches to taxation of resource royalties, etc. Overinvestment in these kinds of industries may also have occurred due to weak environmental policies that failed to internalize fully to these activities their full social cost. Another example of an extremely costly policy biased towards certain industries is the use of unemployment insurance to subsidize the seasonal work typical of some primary industries. Furthermore, since high-technology industries tend to cluster in urban areas whereas many of the more traditional primary and especially resource-based industries operate in remote areas, the Canadian tradition of direct and indirect subsidization of the transportation, telecommunications, and utility costs of remote communities involves a clear sectoral bias. What is required is a systematic examination of the sectoral biases that may inhere in a whole range of existing government policies.

In sum, without attempting anew costly forms of industrial targeting, there are many policy shifts and innovations that offer, in the Canadian context, a real opportunity to address and exploit the relationship between technology and economic growth and to take advantage of the external economies, linkages, and spillovers to which advocates of 'new' industrial policy frequently refer.

NOTES

1 Throughout this chapter I have drawn on recent unpublished work with Marsha Chandler, 'Industrial Policy in Canada and the United States'.

2 D. Crane, 'Toward a Public–Private Partnership: The Role of Government in Promoting Economic Growth' (1995) *Policy Options* 3 at 4–5.

3 See M. J. Trebilcock, M. Chandler and R. Howse, *Trade and Transitions: A Comparative Analysis of Adjustment Policies* (London: Routledge, 1990), at 22–4, criticizing the proposals for targeted industrial policies in high-technology sectors advocated by R. Harris, *Trade, Industrial Policy and International Competition*, Vol. 13, *Studies for Royal Commission on the Economic Union and Development Prospects for Canada* (Toronto: University of Toronto Press, 1985).

4 P. R. Krugman, 'The Current Case for Industrial Policy' in D. Salvatore, ed., *Protectionism and World Welfare* (Cambridge: Cambridge University Press, 1993).

5 Ibid. at 176.

6 Michael Porter, *The Competitive Advantage of Nations*; S. S. Cohen and J. Zysman, *Manufacturing Matters*.

7 M. Itoh *et al.*, *Economic Analysis of Industrial Policy* (Academic Press: San Diego, 1991) at 223.

8 As noted with respect to education by M. J. Trebilcock, *The Limits of Freedom of Contract* (Cambridge, MA: Harvard University Press, 1993) at 257.

9 See D. I. Okimoto, *Between MITI and the Market: Japanese Industrial Policy High Technology* (Stanford: Stanford University Press, 1989), pp. 67ff.; see also H.-J. Chang, *The Political Economy of Industrial Policy* (New York: St Martin's Press, 1994). Chang notes at 76 that government 'announcements' of possible linkages 'can provide obvious focal points for coordination between complementary investments', without requiring direct government assistance such as subsidies.
10 *Competitive Advantage of Nations, supra* n. 6 at 639.
11 See D. I. Okimoto, *Between MITI and the Market, supra* n. 9 at 71.
12 D. Kahneman, P. Slovic and A. Tversky, eds, *Judgment Under Uncertainty: Heuristics and Biases* (Cambridge: Cambridge University Press, 1982).
13 See D. I. Okimoto, *Between MITI and the Market, supra* n. 9 at ch. 3.
14 T. Hatzichronoglou, 'Indicators of Industrial Competitiveness: Results and Limitations' in J. Niosi, ed., *Technology and National Competitiveness* (Montreal and Kingston: McGill-Queens, 1991) at 190.
15 Ibid.
16 L. Davis, 'Technology Intensity of US, Canadian, and Japanese Manufacturing Output and Exports' in J. Niosi, ed., *Technology and National Competitiveness, supra* n. 14 at 27.
17 *Report of the Premier's Council: Industrial Policy Studies*, vol. III (Toronto: Queen's Printer for Ontario, 1988) at 189.
18 T. Hatzichronoglou, 'Indicators of Industrial Competitiveness,' *supra* n. 14 at 191.
19 Cited in M. Porter *et al.*, *Canada at the Crossroads: The Reality of a New Competitive Environment* (Ottawa: Business Council on National Issues, 1991).
20 See, generally, R. Reich, *The Work of Nations* (New York: Knopf, 1991); see also Economic Council of Canada, *Good Jobs, Bad Jobs; Employment in the Service Economy* (Ottawa: Economic Council of Canada, 1990).
21 For a balanced but inconclusive assessment of this issue, see R. M. Cyert and D. C. Mowery, eds, *Technology and Employment: Innovation and Growth in the US Economy* (Washington, DC: National Academy Press, 1987).
22 J. Niosi, *Canadian Multinationals* (Toronto: Between the Lines, 1985) at 172.
23 See, generally, D. G. McFetridge, 'The Economics of Industrial Structure: An Overview' in D. G. McFetridge, ed., *Canadian Industry in Transition* (Toronto: University of Toronto Press, 1986).
24 C. Beltz, *High-Tech Maneuvers: Industrial Policy Lessons of HDTV* (Washington, DC: AEI, 1991) at 14.
25 See Cohen and Zysman, *Manufacturing Matters, supra* n. 60.
26 A detailed analysis of the failure of some of these ventures can be found in S. F. Borins and L. Brown, *Investments in Failure: Five Government Enterprises that Cost the Canadian Taxpayers* (Toronto: Methuen, 1987).
27 Industry, Science and Technology Canada, *Support for Technology Development: A Summary of Federal Programs and Incentives* (Ottawa: Supply and Services Canada, 1989).
28 *Report of the Premier's Council, supra* n. 17.
29 For a basic articulation of the public choice perspective, see A. Downs, *An Economic Theory of Democracy* (New York: Harper, 1957), and M. J. Trebilcock *et al.*, *The Choice of Governing Instrument* (Ottawa: Economic Council of Canada, 1982).
30 M. S. Gertler, 'Canada in a High-Tech World: Options for Industrial Policy' in D. Drache and M. S. Gertler, eds, *The New Era of Global Competition: State Policy and Market Power* at 386.

31 See, for example, S. Haggard and C. I. Moon, 'The South Korean State in the International Economy: Liberal, Dependent, or Mercantile?' in J. G. Ruggie, ed., *The Antinomies of Interdependence: National Welfare and the International Division of Labour* (New York: Columbia University Press, 1983).
32 In *Trade and Transitions*, *supra* n. 3, at ch. 6 my co-authors and I suggest some kinds of new institutional arrangements that could address these problems.
33 See, in the US context, L. Cohen and R. Noll, *The Technology Pork Barrel* (Washington, DC: Brookings Institution, 1991).
34 Indeed, Krugman goes so far as to suggest that the identification of external economies that warrant industrial policies may depend upon *pre-existing* geographical clusters that reveal linkages between industries and firms that can be further exploited. Krugman, *supra* n. 4 at 176.
35 See A. Tupper, *Public Money in the Private Sector: Industrial Assistance Policy and Canadian Federalism* (Kingston: Institute for Intergovernmental Relations, 1982), and 'Federalism and the Politics of Industrial Policy' in A. Blais, ed., *Industrial Policy* (Toronto: University of Toronto Press, 1986).
36 A. Blais, P. Foucher and R. Young, 'La dynamique de l'aide financière directe du government féderal à l'industrie manufacturière au Canada', 19 Canadian *Journal of Political Science* 29.
37 Porter *et al.*, *Canada at the Crossroads*, *supra* n. 19 at 348.
38 See R. Bruce Doern, 'The Political Administration of Government Reorganization: The Merger of DREE and ITC' (1987) 30 *Cdn. Public Administration* 34.
39 For example, this is a major theme in D. I. Okimoto, *Between MITI and the Market*, *supra* n. 9. See also H. Patrick, ed., *Japan's High Technology Industries: Lessons and Limitations of Industrial Policy* (Seattle: University of Washington Press, 1986).
40 *The Competitive Advantage of Nations*, *supra* n. 6 at 627–30.
41 See M. J. Trebilcock, *The Limits of Freedom of Contract*, *supra* n. 8 at 257.
42 *Making Democracy Work: Civil Traditions in Modern Italy* (Princeton: Princeton University Press, 1993).
43 Ibid. at 161.
44 Ibid. at 170.
45 I have developed this theme elsewhere. See R. Howse, 'Retrenchment, Reform or Revolution: The Shift to Incentives and the Future of the Regulatory State' (1993) 31 *Alberta Law Review* 455.
46 I do not mean that these industries never produce technological spillovers; rather I am suggesting that if overinvestment in them is observed (along with relative underinvestment in industries that appear to generate more spillovers) this may well be due to government failure rather than market failure.

2

THE ROLE OF COOPERATIVE INDUSTRIAL POLICY IN CANADA AND ONTARIO

David A. Wolfe

INTRODUCTION

The current era of economic and technological change is marked by a profound degree of social dislocation and uncertainty. At the root of this change are three interrelated processes: the emergence of a new information technology paradigm that is dramatically altering the economic calculus of production and distribution throughout the industrial economies; the phenomenon of globalization which is intensifying the linkages and interdependence between the economies of Europe, North America and East Asia; and the gradual decline of Fordist methods of standardized mass production and distribution which is altering the occupational structures and patterns of employment throughout these economies. The ability of individual economies and societies to respond to the stress of this social dislocation is determined, in large measure, by the capacity of their existing institutions to adapt to the changes underway. While the nature and extent of the adaptation required may seem novel, the process itself is not. Periods of rapid economic and technological change are characterized by a condition of extreme uncertainty. They place a high premium on the ability to acquire, absorb and diffuse relevant knowledge and information throughout the various institutions that influence the process of economic development.

A number of writers have remarked on the critical role of institutional supports for the process of technological change. Bengt-Åke Lundvall, Richard Nelson and Christopher Freeman all focus on how effectively the institutions supporting innovation cohere into national systems of innovation. A similar body of work, linked to the literature in economic geography, specifies the increasingly valuable role that institutional infrastructures at the regional and local level play in supporting the economic dynamism of some of these localities. This chapter draws upon these perspectives to assess the contribution of public policy to the Canadian innovation system over the past decade. It examines the record for industrial and technology policy at both the federal level and in Canada's largest province. It concludes by evaluating the extent to which both levels of government have charted a new direction for Canadian industrial policy that reflects the theoretical insights of the literature cited above by supporting the development of new institutions that can foster adjustment to the emerging technological paradigm.

PARADIGM SHIFTS AND NATIONAL SYSTEMS OF INNOVATION

Recent work in the neo-Schumpeterian tradition has accentuated the significance of the current shift in the techno-economic paradigm. The concept of a techno-economic paradigm involves changes in the underlying technology system which are so far-reaching in their effects that they exert a transformative influence on every aspect of behaviour throughout the industrial economy. Technological changes of this magnitude alter not only the competitive structure of specific branches or sectors of the economy, but the entire input cost structure and conditions of production and distribution throughout the economy. The emergence of a new techno-economic paradigm is distinguished from less pervasive forms of technological innovation by the specific set of characteristics associated with its most important input or key factor: a relatively low and constantly falling cost curve, plentiful supply, and ease of application across many sectors of the economy. The new key factor does not diffuse throughout a modern economy as an isolated input, but rather at the core of a rapidly growing system of technical, social and managerial innovations (Freeman and Perez, 1988, 58–61).

At the heart of the new techno-economic paradigm is the convergence of an integrated set of computer, communications and video technologies that are based on semiconductors and share in common the capacity to process and transmit data in digital form. The current diffusion of these technologies throughout the industrial economies may have effects as vast and sweeping as those that accompanied the second industrial revolution at the turn of the century (Organization for Economic Co-operation and Development, 1988, 35; Tapscott, 1996). While the microelectronics revolution is increasing the capacity to handle information at an exponential rate, knowledge-based inputs are simultaneously becoming a more salient component of every aspect of production. Some analysts have argued that this development signifies the shift to a new form of 'post-industrial'

(Bell, 1973) or 'post-capitalist' (Drucker, 1993) society. In effect, the new paradigm mobilizes knowledge, social intelligence, and innovative capacity in a dynamic way within the context of a capitalist mode of production. If knowledge is understood to include not just R&D, but also design, engineering, advertising, marketing and management, then it is clear that knowledge-based inputs are becoming the defining feature of both manufacturing and service industries in the contemporary economy. These developments lead to the characterization of the emerging period more appropriately as one of 'knowledge-intensive' capitalism (Mytelka, 1987; Florida, 1995).

The increasing salience of knowledge and innovation in the new paradigm has also drawn attention to the innovative capacity of national economies in responding to this change. Periods of rapid technological change are characterized by a condition of extreme economic uncertainty in which the frontier of technological possibilities is in a state of flux. Past technological paradigms or trajectories no longer serve as adequate guideposts in the emerging economic and technological environment. Studies of the innovation process reveal that simple technical superiority is no guarantee of economic success. Such periods place a premium on the ability to acquire, absorb and diffuse relevant scientific and technological information throughout relevant educational, economic, or political institutions. As a consequence, these periods accentuate the importance of the collective ability of a nation's institutions to search for, and apply, relevant pieces of knowledge. The extent to which a nation's capacity for scientific learning and technological adaptation is supported or weakened by its institutional structure is critical to its success in these periods.

The differing ability of individual nations to take advantage of new opportunities associated with the emerging techno-economic paradigm has led a number of writers to adopt the concept of national systems of innovation (NSI) to analyse these differences. The concept of a system of innovation describes the intersection of institutional structures and social and economic forces that determine the capacity of individual societies to adapt to the kinds of changes discussed here. The first use of the concept is attributed by Christopher Freeman to Bengt-Åke Lundvall in his contribution to the edited volume on *Technical Change and Economic Theory* (Lundvall, 1988). However, the first published discussion of the concept is found in Freeman's own study of technological change in the Japanese economy. Both Freeman and Lundvall trace the roots of the NSI approach back to Friedrich List's classic study of the National System of Political Economy in 1841. List's study, devoted to the issue of how Germany could overcome Britain's lead in industrialization, was concerned with the question of learning about, and applying, new technology in the context of an industrializing economy. According to Freeman,

> Not only did List analyze many features of the national system of innovation which are at the heart of contemporary studies (education and training institutions, science, technical institutes, user–producer interactive learning, knowledge accumulation, adapting imported technology, promotion of strategic industries, etc.), he also put great emphasis on the role of the state in coordinating and carrying through long-term policies for industry and the economy.
>
> (Freeman, 1995, 7)

In Freeman's own study, he defined national systems of innovation as 'the network of institutions in the public and private sectors whose interactions initiate, import, modify and diffuse new technologies' (Freeman, 1987). Freeman emphasizes the role that social and political institutions play in supporting the adoption and dissemination of scientific and technical knowledge. In his analysis of the Japanese system of innovation, he analyses the contribution of four components: the role of government policy; the role of corporate R&D the role of the education and training system; and the general structure of industry.

Lundvall provides a somewhat more inclusive definition of the concept. One of the crucial differences between his approach and that of Freeman is the importance he attaches to the patterns of interaction between firms as part of a collective learning process in the acquisition and use of new technical knowledge. This flows from his belief that innovation is increasingly tied to a process of interactive learning and collective entrepreneurship, especially in terms of the relationship between producers and users of new technology. For him,

> a system of innovation is constituted by elements and relationships which interact in the production, diffusion and use of new, and economically useful, knowledge and . . . a national system encompasses elements and relationships, either located within or rooted inside the borders of the nation state.
>
> (Lundvall, 1992a, 2)

The main elements of the system in his conception include: the internal organization of firms; the network of interfirm relationships; the role of the public sector; the institutional set-up of the financial sector; and the degree of R&D intensity and the nature of R&D organization.[1]

The concept of the national system of innovation comprises a much broader range of institutions and activities than are usually subsumed within the framework of industrial and technology policy, yet the contribution of the public sector is central for Freeman, Lundvall and Nelson. Although most innovative activity in national economies occurs within the boundaries of the private firm, the role of the public sector in stimulating and sustaining innovative behaviour is critical. In most of the industrial countries, the government-supported share of funding on research and development ranges from a low of 20 per cent to more than 50 per cent. Moreover, the public sector maintains a vast array of infrastructural supports critical for the innovation process in the form of the post-secondary educational system, public R&D facilities and a wide range of institutions that support the diffusion of technology. The concept of a national system of innovation thus draws attention to the effectiveness of public policies in fostering the kinds of private sector activity necessary to develop and maintain a sustained innovative capacity (Niosi, *et al.*, 1993, 212).

THE ROLE OF INDUSTRIAL AND TECHNOLOGY POLICY IN CANADA

The principal study of Canada's national system of innovation is situated firmly within this body of literature, yet focuses primarily on the 'industrial' system of innovation. It identifies the constituent parts of Canada's system of innovation as those organizations which contribute to the generation of knowledge, including R&D performing firms, universities, other nonprofit organizations and government laboratories, as well as those that apply knowledge and the interactions between the two (McFetridge, 1993, 299). To the extent that it considers the role of industrial and technology policy, it focuses on a limited number of institutions, such as the National Research Council, or policies, such as the Defence Industry Productivity Program. While these are among the more prominent examples of federal initiatives, they encompass far from the full panoply of federal initiatives in this field.

Since the election of the federal Conservatives in 1984, the field has experienced a continuous process of change and renovation. While the Conservatives are generally viewed as pursuing a neo-liberal trade and industrial adjustment strategy, symbolized by the Free Trade Agreement with the US, this did not hinder them from implementing a number of key initiatives in the field of technology and industrial policy. In 1987, the federal and provincial governments jointly signed the first National Science and Technology (S&T) Policy, which created the Council of Science and Technology Ministers. Its objectives, among others, were to optimize the existing S&T policies of both orders of government and to identify S&T policies and program actions that might be undertaken jointly.

In 1987, the federal government also launched its own science and technology strategy, 'InnovAction'. The official goals of the program were to increase industrial innovation and technology transfer, develop and promote strategic technologies, manage federal science and technology resources more effectively, ensure the adequate development of human resources needed for science and technology and promote a more science-oriented culture in Canada. The strategy was assigned a five-year budget of $1.3 billion allocated among the following program initiatives: Science and Technology Programs (including the new Strategic Technologies Program), $466 million; the Canadian Space Program, $256 million; funding for fifteen National Networks of Centres of Excellence, $240 million; increases to the three federal research granting councils of $200 million; and a new Canadian scholarships program for students in science and technology, $80 million. The overall program thrust was refined somewhat in the early 1990s and the financing extended for a further five years (Dufour and de la Mothe, 1993b, 35–6; 1993a).

The most interesting initiatives to flow from the strategy were two programs, the Strategic Technologies Program (STP) launched in 1988 and the Microelectronics and Systems Development Program (MSDP) effectively in operation since 1989. The STP, a five-year, $100 million program, was designed to enhance the international competitiveness of Canadian industry through the development, acquisition, application and diffusion of three enabling technologies: information technology, biotechnology and advanced industrial materials. The explicit goal of the program was to promote more effective vertical and horizontal networking among firms; eligible recipients were private sector corporations and partnerships in Canada representing alliances from the private and institutional sectors. Its funding was subdivided into two major streams: one to support long-term collaborative and pre-competitive R&D between firms and other public research organizations; the second supported downstream development in the innovation process by funding technology feasibility trials, demonstration projects, international market development, standards development and research to meet regulatory requirements (Botham and Giguère, 1993, 82–3).

Although the program was intended to run for five years, an evaluation study conducted near the end of its original lifespan concluded that its overall design was strong and that positive immediate effects could be attributed to the program. In addition, the study found that each dollar of STP assistance invested under the program attracted an additional $2.50 in matching funds from other sources. The study made a number of recommendations to improve the delivery of the program and recommended that it be extended. During the peak years of the program, from 1991–2 to 1995–6, total spending averaged $20 million a year, but it has since fallen to $15 million a year in the aftermath of the spending restraint introduced in the 1995. The unique and noteworthy feature of this program was its strong emphasis on the need for collaborative relations among a network of firms as a criterion for receiving funding. Clearly modelled on some of the cooperative technology ventures launched in the US in the 1980s, such as the Microelectronics and Computer Corporation (Gibson and Rogers, 1994) and SEMATECH (Grindley *et al.*, 1996), the STP program attempted to build the features of networking and cooperation into its project design and structure.[2]

The MSDP Program was announced around the same time as a five-year $60 million initiative designed to support the development of advanced microelectronics and information technology systems and to enhance the growth and international competitiveness of companies in the manufacturing, processing, or service industries. Companies with capabilities in systems engineering and/or integration, and developers of microelectronics or speciality electronics for integrated systems applications, were eligible to apply. This program was also subdivided into two components: one to support R&D in microelectronic components; and the other for R&D in systems based on advanced microelectronic and information technologies (Botham and Giguère, 1993, 82–3). In contrast to the STP program, spending on the MSDP was wound down after 1995, with a total of $68.5 million expended during the course of its existence.

In addition to these new initiatives launched in the late 1980s, a number of older federal programs continued to play an important role in supporting R&D. The major federal spending program in this area was the Defence Industry Productivity Program (DIPP). Created in 1959, on the heels of the Joint Defence Production Agreement between Canada and the US, DIPP was to provide support for Canadian firms in their competition with their larger US counterparts for a share of the lucrative continental market in defence contracts. Modified in 1968 and placed under the control of the federal Department of Industry, the program has provided strong support for the Canadian aerospace and defence-related industries, overwhelmingly located in Ontario and Quebec. Under the terms of DIPP, firms were able to obtain conditional grants for the eligible costs of R&D projects, to cover the costs of acquiring advanced production equipment for modernizing or upgrading manufacturing facilities, to cover the costs of establishing qualified Canadian suppliers of defence-related products (source establishment assistance) and to carry out market feasibility studies. DIPP funding has averaged between $150 and $300 million annually, but was continuously the main source of support for R&D and innovation in the aerospace industry (Botham and Giguère, 1993, 82). Despite its success and the critical role it played in supporting innovation in key sectors such as aerospace, the program was terminated in the 1995 budget as part of the general reduction of business subsidies. This action sparked a firestorm of protest from key sectors of industry and it was subsequently replaced by the new Technology Partnerships program.

Beside the industry support programs cited above, the federal government provides an extensive set of infrastructural supports for the national innovation system in Canada. One of the principal contributions made by the federal government is through its premier science and engineering organization, the National Research Council (NRC). The NRC has 3,000 staff located in ten centres across the country and accounts for 13 per cent of total spending on R&D by federal departments and agencies. Its three primary objectives are: to undertake world-class research, build partnerships with industry and other research organizations, and concentrate on areas of research that raise Canada's competitiveness (Dufour and de la Mothe, 1993, 32). Despite the size and overall contribution of the NRC to the conduct of basic science, its role within the system of innovation has been the object of a wide range of criticism over the past decade and a half and the subject of numerous reports and evaluations. A key problem with federally performed R&D has been the lack of communication between government laboratories and their clients. In reaction to these criticisms and in response to the recommendations of some of these reports, the NRC terminated some of its programs and reoriented the focus of others. It realigned its institutes into five technology groups chosen on the basis of their importance for the emerging knowledge-based economy. In addition, it worked to develop a more entrepreneurial culture and strategy that concentrates on the role of regionally based innovation systems across the country. This shift accentuates the potential contribution of the NRC to the regionally based innovation systems discussed in more detail below (Lyrette, n.d. 7).

Complementary to its role in producing basic science and engineering research, the public infrastructure also plays a critical role in supporting technology adoption and diffusion. The principal federal program in this area is the Industrial Research Assistance Program (IRAP), run by the NRC. IRAP was established in 1962 to provide technological assistance to Canadian firms to help them solve technological problems in a timely and cost effective manner. It maintains a national network of 245 Industrial Technology Advisers (ITAs) composed of staff from the NRC and some one hundred other organizations across the country, including universities, federal government organizations and provincial research organizations. The ITAs work with firms to help them examine the nature of their technology problems and identify possible sources to help solve their problems, such as the NRC itself, other public research organizations, universities or private companies. IRAP also makes grants available to the firms to acquire the necessary technology or hire staff to implement the recommended solution. Although consistently praised as one of the most effective public policy instruments in promoting the adoption and use of technology in Canadian industry, IRAP suffered from a series of budgetary cutbacks in the late 1980s and early 1990s, but its funding has been stabilized in recent years and was given a new lease in the 1997 budget.

During the early 1990s, the federal government also invested considerable resources in working with industry associations as part of its sector competitiveness campaigns. This initiative combined some elements of the experience gained in working with sectoral associations in an advisory capacity in the free trade initiative (SAGITs), with some of the experience gained in the ill-fated Prosperity Initiative launched in the closing years of the Conservative government. The traditional strength of Industry Canada's sector branches has been the ability to produce competitiveness framework studies based on a benchmarking of Canadian sectors with those in other countries (Doern and Tomlin, 1996, 178–9). The goal of the sector competitiveness campaigns was to improve the competitive position of selected sectors where strategic analysis indicated that focused initiatives can yield significant benefits. A key design feature of these campaigns was the notion of joint action involving shared risk and commitment between the industries involved and government departments. In total, four of these frameworks were completed in 1994–5, another ten were nearing completion in 1995–6, with a further fifteen slated for completion by the end of 1996–7.

The role of the federal government in science and technology policy has been subject to considerable strain since the election of the Liberal government in 1993. The election of the new government generated high expectations based on the party's policy platform released at the outset of the 1993 campaign, the famous Red Book, which highlighted the importance of innovation in the new ideas-based economy. It highlighted the dynamic role of small and medium-sized enterprises in a growing economy, the need to revitalize the manufacturing, resource and service industries, and to enhance the ideas-based sectors of the economy, and the importance of supporting the communities in which these businesses are grounded. It emphasized the need to move research results from the lab to the marketplace more effectively and to help Canadian business adopt and use new technology more effectively. A main theme of the strategy was the need to build clusters of economic activity within different regions of the country in order to create value-added jobs. The principal means identified for realizing this goal were fostering an entrepreneurial spirit and developing forums for economic cooperation and networking. In both cases, the focus of the Red Book was on the need for greater attention to the subnational level (Paquet and Roy, 1995, 137–8).

In the autumn of 1994, the Liberal government released another series of policy documents outlining its broader economic agenda. The framework document issued by the Department of Finance stressed the role that it saw for innovation policy as one of the four pillars of its economic agenda. A companion document issued by Industry Canada, *Building a More Innovative Economy* (Orange Book), placed the government's S&T initiatives within the context of its broader economic agenda. Policy initiatives were to be pursued in four key areas: trade, infrastructure, technology and the climate of the marketplace. Of the four key areas, the two most central to the concerns of this chapter are infrastructure and technology. In the area of infrastructure spending, the government indicated its strong intention to support the information highway. It highlighted its commitment to regulatory reform in the area of telecommunications policy, its efforts to extend the CANARIE network with capital spending, and its support for the linking of all schools and libraries in the country to the Internet through the SchoolNET program. With respect to technology, it outlined some of the key issues to be addressed in the formal S&T Review (launched as part of a more comprehensive process of program review), including the need for a more systematic approach to the commercialization of R&D, the need to develop a strong scientific culture in Canada, the need to establish which scientific and technological developments Canada should pursue, the need to ensure that federal labs play an effective role in the commercialization of technology and the need for measures to promote the rapid diffusion of technology to industry (Government of Canada, 1994). The Orange Book reiterated the campaign commitment to creating partnerships, but translated it from a process of fostering community-based networking and cooperation into one designed to help small business by cutting red tape, increasing guarantees for loans to small business and providing better support for exports (Paquet and Roy, 1995, 146).

The federal budget for 1994 reflected some of the broad themes outlined in the policy documents, but clearly further development awaited the outcome of the S&T Review. Chief among the measures announced were funding for the Canadian Technology Network, additional funds for the National Research Council, along with a stabilization of funding for the Networks of Centres of Excellence and the Research Granting Councils. The announcement of a new Canadian Technology Network (CTN) in the federal budget of February, 1994 suggested a dramatic expansion of IRAP-type services. The CTN, established in 1995, was to build on the success of IRAP and its network of Industrial Technology Advisers. Its objective was to provide firms with quick access to all of the technology and related business expertise currently available from a variety of different sources. The CTN provides firms with access to a suite of different services, including access to relevant government and private sector services and programs, benchmarking, managing technology, training, financing, standards and regulations, and assessment and evaluation of technologies. Each CTN member has specialities in certain sectors, technologies or business expertise; they also act as a networking agent, bringing together parties with the appropriate skills and expertise to address their concerns (Lyrette, n.d. 10).

The first budget seemed to provide some degree of support for the directions promised in the Red Book. However, this did not last long. In the next budget of February, 1995, the government signalled a change in direction away from the themes of the Red Book and towards a priority on deficit reduction. Not awaiting the outcome of the S&T Review, it levelled a series of major cuts at program spending. The cuts were substantial across the board and hardly any portfolio was spared, including the S&T one. Industry Canada was particularly hard hit, with expenditure reductions of more than 42 per cent planned over a two-year period, including most of its industrial subsidies, such as DIPP.

The last remaining vestige of hope for the original promise of the Red Book rested with the S&T Review. The federal S&T Review has been the subject of much discussion and some criticism, both during the period of its existence and since the release of its report. Part of the reason for the intense scrutiny was the high expectations that it engendered in the science and technology research community. The consultation itself was very inclusive, involving twenty-nine local meetings, five regional meetings and a final national meeting with a total of twenty-five hundred participants. The inclusiveness of the process sustained the belief that the 1995 budget was a temporary step backwards and that the government intended a change in policy direction towards the goals articulated in its campaign platform of 1993 (de la Mothe, 1996, 415–16). In addition to the numerous consultations held by the Secretariat, there were four interdepartmental committees at work, each of which produced an internal report, as well as a report issued by the National Advisory Board on Science and Technology. After a long travail of nearly two years, the results were finally released in March 1996. The strategy established three related goals for building a dynamic and forward-looking Canadian innovation system: (1) Canada should be among the world's leaders in 'applying and commercializing S&T for sustainable job creation and economic growth', (2) that S&T be applied in such a way as to maximize the quality of life of Canadians, (3) that Canada retain and enhance its capabilities to advance knowledge in all major areas of scientific and technological endeavour. To be precise, the goal was

> to create in Canada world centres of excellence in scientific discovery; to build a broad base of scientific enquiry; to foster Canadian participation in all major fields of science and technology; and to ensure that new knowledge can be acquired and disseminated widely, from Canadian sources and from around the world.
>
> (Government of Canada, 1996, 6)

Having identified these goals, the strategy noted the overriding need for the federal government to establish clear priorities for spending in light of the continuing pressure to reduce its fiscal deficit. As a result of the reductions already underway, the report underlined the need for public spending to focus on core activities in the S&T policy area and to find more efficient and effective ways to deliver those activities. The principal means identified for improving on the efficiency of delivery mechanisms was the increased use of partnership arrangements between government departments and agencies and other key components of the innovation system. The strategy identified the core S&T activities of the federal government as: (1) funding that research which supports the mandates of federal agencies, (2) providing research support to universities, the Centres of Excellence, and other non-governmental research institutes, (3) supporting private sector research and development, and (4) disseminating knowledge, building information networks and acting as an information analyst (p. 10). Despite the positive-sounding rhetoric of the strategy document, the action plans associated with it fell short of the expectations that had been engendered. In the eyes of one commentator, at least,

> it took too long to deliver and resulted in a feeling that the policy leadership developed by the government was lost. It was not as comprehensive as had been promised . . . [and] . . . suffered from the fact that 'new' programs, like the Technology Partnerships Canada program, simply appeared to be replacements for old programs, like the Defence Industry Productivity program . . .
>
> (de la Mothe, 1996, 415–16)

In conjunction with the release of the federal strategy, the major budgetary development in 1996 was the announcement of the new Technology Partnerships Canada program to provide support to private sector partners, such as those in the aerospace industry, in their efforts to commercialize high technology products and processes. This was the industry hardest hit by the cancellation of the DIPP in 1995, and although the Technology Partnerships program differs in important ways from its predecessor, it went a long way towards satisfying the concerns of the industry. The most recent budget of the Liberal government, brought down in February 1997, contained several significant announcements for the science and technology portfolio. Chief among these was the establishment of the Canada Foundation for Innovation (CFI) with an initial allocation of $800 million over a period of five years. The CFI will provide funds on a matching basis to the provinces or industry and for the modernization of research facilities in the natural sciences, engineering and health sciences at universities, colleges, research hospitals and non-profit research institutions. In addition to the CFI, the 1997 budget also made the Networks of Centres of Excellence program permanent and stabilized its annual funding at $47.4 million – but this was largely achieved by reallocating money from the budgets of Industry Canada and the granting councils. Finally, the funding for the popular and successful IRAP run by the NRC was also stabilized.

While the 1997 budget certainly provided better news for science and technology policy, the initiatives should not be regarded as a panacea for all of the problems that remain. This brief review of the direction of federal policies over the past decade under both Conservative and Liberal governments depicts a story of imperceptible progress through a series of fits and starts. On the positive side, both governments recognized the need to improve the dynamism of the Canadian system of innovation by enhancing the performance of key elements, such as the basic research infrastructure and the supports afforded to both SMEs and larger firms. Most significant has been the growing emphasis on the importance of partnership in the design of a key series of programs over the past decade and in the desirability of increased networking and cooperation among firms in key sectors and between individual firms and the public research infrastructure. On a less positive note, however, is the repeated inconsistency in both key funding and policy initiatives. This has created a high degree of confusion, at best, or scepticism, more likely, in the eyes of the private sector partners with whom the federal government must interact.

THE EXPANDING ROLE OF THE REGION

While most of the work in the neo-Schumpeterian tradition on systems of innovation has focused on the national level of analysis, an expanding stream of work has drawn attention to the role of the regional or subnational units of government in supporting and sustaining the innovation process. Complex systems of technology, production processes, industrial organization, and their supporting infrastructures of social and political institutions, frequently exhibit distinctive spatial characteristics. Patterns of production relations tend to aggregate over time among interrelated networks of firms drawing upon the distinctive skills and characteristics of local labour markets in specific regions. Similarly, periods of transition from one set of technologies (and its associated production paradigm) to another, tend to coincide with a reorganization of the spatial distribution of production. These periods of transition pose distinctive challenges for nations, and the regions within them, in terms of their ability to adapt to the reconstitution of the spatial bases of production. Indeed, recent analyses have begun to adopt the concept of national systems of innovation for use at the subnational level as well (Cooke, 1998).

The emergence of the new paradigm thus underlines the changing relationship between national and subnational, or regional, levels of government. Charles Sabel links this phenomenon to the transition from the old

Fordist mode of standardized mass production to the emerging post-Fordist mode of flexible, knowledge-intensive production. At root, 'the relationship between the economy and its territory is changing'. The enhanced role for subnational levels of government stems, in part, from the significance attached to geographically based local networks of firms and knowledge-creating forces. The emerging knowledge-intensive industries in the new paradigm exhibit a marked tendency to cluster around regional growth poles – not only are the firms in these industries closely linked to each other but they also tend to locate close by their clients. Further, the service industries that provide inputs to them also tend to be located closely – in this sense, industrial geographers who study this phenomenon refer to regional agglomerations of firms. The evolving post-Fordist economy is witnessing the formation and revitalization of vibrant industrial districts in these new regional economies – ranging from the Third Italy to the other members of the four motors of Europe, such as Baden-Württemburg, a host of other regional districts in Europe and even some of the innovative experiments conducted at the state level in the US during the 1980s (Sabel, 1989; Murray, 1991, 19–21).

Industrial geographers maintain that these new industrial spaces involve agglomerations composed of collectives of interdependent producers whose interrelations induce them to converge geographically (Scott and Storper, 1992). These local and regional agglomeration tendencies are reinforced by the information-intensive nature of the technological activities associated with the emerging techno-economic paradigm. The value of face to face interaction, particularly in the context of an abundant technical and professional labour force and a supportive regional institutional infrastructure is inestimable. The renewed emphasis on the link between geographic proximity and technological dynamism is rooted in recent insights into the nature of the innovation process itself. Technologies tend to develop along pathways or trajectories that are characterized by strong irreversibilities. These trajectories are reinforced as certain choices are made and others are foreclosed; this results from the fact that technologies are the product of interdependent choices, or 'network externalities'. They are given to a variety of user–producer and user–user interactions; as the number of users of a given technology rises it reduces the possibilities for different patterns of use by others (Arthur, 1994).

One reason why technologies follow these pathways is because of the effects of technological spillovers in the economy. The knowledge of how to do certain things technologically frequently derives from the knowledge of how to do other things, and it, in turn, contributes to the knowledge of how to do related things. Sometimes these spillovers follow the lines of traded input–output relationships in the economy, but often they occur along lines that are not traded. These technological spillovers are tied to knowledge and practices that are not always codified or explicit. They are frequently shared among firms or transferred from firm to firm through various forms of networks, such as user–producer relationships, strategic alliances, R&D consortia, collaborative training and marketing schemes, and supportive public infrastructure such as educational institutions. They may also derive from other institutional arrangements, such as the particular norms and rules governing the functioning of local labour markets. These forms of collaboration and networking give rise to the existence of untraded interdependencies within a regional economy (Dosi, 1988, 1142–7; Storper, 1995).

While the concept of untraded interdependencies is grounded in the literature of evolutionary economics and economic geography, it is not a narrowly economic concept. Economic practices are rooted in a broader array of social and political institutions, both private and public, that provide the necessary support for their effective operation. Put another way, a critical variable in the relative capacity of different regions to maintain the conditions that support a dynamic technological trajectory, is the capacity for self-organization on the part of the regions' public and private institutions. This capacity for self-organization has recently been linked to the existence of the stock of social capital in the region. Social capital refers to various features of the social organization of a region, such as the presence of shared norms and trust that facilitate coordination and cooperation among individuals, firms and sectors for their mutual advantage. It secures the conditions that enhance the benefits derived from more tangible investments in physical and human capital. Without its supportive functioning, high levels of these more tangible forms of investment may fail to produce the benefits that should potentially result from them (Putnam, 1993, 167–76).

The concept of untraded interdependencies and the existence of social capital are closely linked to a capacity for networking. The capacity for networking is increasingly seen as essential for tapping into the shared intelligence of both the individual firm or organization, as well as a collectivity of firms within a given geographic space. The spatial or regional dimension of networking refers to the propensity to share information, exchange knowledge and pool resources within a regional economy based on the existence of a high degree of trust. More and more cases can be found of emerging cooperative relationships between networks of producers – between large assemblers and smaller suppliers in the auto industry, between networks of small producers, such as exist in the Emilia-Romagna industrial district of Italy, and even among large producers in the computer and telecommunications industries that make up the core of the new information technologies. The growing costs of R&D, as well as the increasing complexity and knowledge-intensity of new scientific research and product development make the challenge more forbidding for individual firms – hence the growing importance of regional agglomerations of innovative firms in the collective advance of technical knowledge. The key elements of a networked regional economy include a dense network of public and private industrial support institutions, high-grade labour market intelligence and related vocational training mechanisms, rapid diffusion of technology transfer, a high degree of interfirm networking and receptive firms well-disposed towards innovation. The very density of these networks and institutional supports is often interpreted as a sign of the vibrancy of a regional economy (Cooke and Morgan, 1993, 562).

The consequences of regional agglomeration for firms in both the emerging information-intensive and the mature restructuring sectors of the economy draw attention to the role played by subnational or regional levels of governance. Regional economies consist of more than just individual behaviours of firms, or even networks of firms, and their employees. They are also constituted by the cultural traditions and institutional structures that facilitate and regulate economic behaviour and social activity. The expanding role of subnational governments in creating and maintaining these conditions is significant enough for one observer to comment that, 'On the global economic map the lines that now matter are those defining what may be called "region states".' For Ohmae, 'region states' are natural economic zones that tend to have strong linkages with the emerging global economy and be of adequate size to maintain the communications and transportation infrastructure and supporting professional services needed to sustain the regional agglomerations of firms described above (Ohmae, 1993, 78–80).[3]

Given that the region is becoming an increasingly important locus of activity and competitive advantage in the emerging global economy, it is not surprising to find that regional levels of government, some of which fit Ohmae's conception of 'region states', have begun to adopt a variety of policy instruments to foster the kind of interfirm networking and regional innovation patterns found in the more successful economies in the new paradigm. Innovation strategies developed at the subnational level target government support through a number of specific policy instruments, including the provision of technology infrastructure support; the creation of sector-wide consortia for the collective provision of basic services, such as accounting, design, payroll preparation and marketing information; the promotion of joint ventures, strategic alliances and a strategic role for trade associations to foster the creation of interfirm synergies; the creation and upgrading of labour force skills through apprenticeship programs, vocational training schemes and the establishment of specialized training institutes; and, finally, the creation of specialized, regionally or sectorally based development agencies (Scott, 1992, 225–6; Murray, 1991, 70–2).[4]

In a growing number of these instances, the public sector is assuming the role of social animator or facilitator of the desired forms of private activity,

> The elaboration of regionalized or localized public–private interfaces adds substantially to the performance of regional economies by enabling SMEs (small and medium enterprises) to meet more effectively the requirements of their customers in respect of technology, quality and training. Regional economies displaying dense inter-firm and public–private interactions of this kind may be expected to show better than average growth performance.
>
> (Cooke and Morgan, 1993: 554)

As such, they represent a new innovative form of regional policy-making, characteristic of what has been termed the *intelligent region*, which is *institutionally reflexive* in its nature. The governments of these intelligent regions are fundamentally learning organizations, constantly concerned with enhancing their capability to evaluate and refine their policies for promoting regional innovation patterns (Cooke, 1997).

THE ROOTS OF INDUSTRIAL POLICY IN ONTARIO

Industrial policy in Ontario has traditionally been characterized more by a state of benign neglect than conscious direction, although there have been periods of notable exception. In the expansionary atmosphere of the 'golden age' after the Second World War, provincial governments in Ontario were content to rely upon federal macroeconomic management of the economy and Canada's improved trade relations to create the appropriate context for provincial economic development. To the extent that they pursued any consistent set of economic development policies, they focused on the supply side rather than the demand side. Provincial governments in Ontario viewed their primary role as the provision of infrastructure investment in highways, electric power generating capacity, support for municipalities in building water and sewage systems, and, in the decade of the 1960s, a dramatic expansion of the post-secondary educational system. In policy areas of critical import for the provincial economy, such as the imbalance in automotive trade between Canada and the US, the provincial role was constrained by its relative exclusion from jurisdiction over trade issues. The Automotive Products agreement negotiated by the federal government with the US in 1965 was the most significant piece of industrial policy for the province in this period. Overall, the Conservative governments which ruled from 1943 to 1985 were preoccupied with maintaining a 'favourable investment climate' for private business in the province by controlling spending and managing the provincial debt (Rea, 1985, 22–3, 204–5).

This complacency began to fade in the late 1970s as the industrial heartland experienced the economic slowdown and restructuring that affected other industrial economies in the wake of the oil price shock and mid-1970s recession. In this environment of increased economic uncertainty, the Conservative government in Ontario began to gravitate towards a new set of industrial policies that initially tried to cushion the impact of the adjustment process on the provincial economy. In the late 1970s and early 1980s, the provincial government also came under pressure to provide financial assistance to firms facing the prospect of industrial restructuring, especially in the automotive industry. The North American industry experienced a massive process of restructuring in the late 1970s, involving substantial capital outlays. Faced with intense competition from Japanese manufacturers and the possibility of global relocation, North American manufacturers looked to the federal and provincial governments to provide financial assistance.

This experience led three successive governments in Ontario to adopt a more coherent framework for industrial and technology policy. The first phase occurred under the Conservative government with the formation of the Board of Industrial Leadership and Development (BILD) in January 1981. The board, a cabinet committee, was charged with responsibility for coordinating all government spending on industrial, resource, transportation and regional development, budgeted at $2 billion in 1980/81. In addition, it was assigned a new series of economic initiatives which were to amount to $750 million over the five years from 1980. The larger portion of the funds allocated to the BILD initiative amounted to a repackaging of existing funds. However, by the 1983 budget, the government had raised the planned allocation to $1.1 billion, of which $503 million was designated for science and technology projects. The new portion of these funds represented a notable shift in the focus of provincial industrial and technology policy. With the BILD program, the policy concern shifted downstream to encourage the process of innovation through the commercialization of research results, as well as the adoption of new process technologies by existing industries.

A key element of the BILD program was the commitment of $100 million over five years to fund five industry-oriented technology centres: Automotive Parts Technology, Resource Machinery, Farm Equipment and Food Processing, Advanced Manufacturing (two separate CAD/CAM and Robotics centres), and Microelectronics. The mission of the centres was to enhance the competitiveness of Ontario industry and support the diffusion of new technology by aiding in the development of high technology production through advice, funding and prototype development; providing training programs in advanced technologies to industry; providing information, seminars and technology awareness activities; and demonstrating technology hardware and software applications. In part, the establishment of the centres responded to concerns raised in the report of a provincial task force on microelectronics which noted the slow pace of adoption of new process technologies. Some of the remainder of the funds went to help establish a new biotechnology firm, Allelix Ltd; to support the diffusion to Telidon technology which had been developed by the federal Department of Communications; to establish a research institute in hydrogen systems at the University of Toronto; and to continue support for the Ontario Research Foundation (Jenkin, 1983, 73; Miller, 1983, 18–19; Grossman, 1984).

The BILD initiative represented the first halting step towards shaking off the complacency of Ontario's industrial policy and confronting the reality of an emerging techno-economic paradigm. The technology centres achieved a mixed degree of success. Some, such as the Automotive Parts Centre in St. Catherines, were credited with having helped Ontario's parts manufacturers successfully to adopt the new technology of statistical process control in the early 1980s and to position themselves to participate in the tremendous boom that buoyed up the provincial auto industry in the latter part of the decade. However, a condition of their establishment had been that they achieve 50 per cent self-sufficiency by their fourth year and a review of their success was required by 1987. An evaluation conducted for the new Liberal government in 1986 concluded that they were achieving a high benefit/cost ratio and were playing a 'critical role' in promoting economic development. None the less, the government instructed the centres to submit new business plans with provision for higher levels of cost recovery, streamlined operations and reduced expenditures. In the end the government sold or privatized most of the centres, and by 1990 all but one had ceased operation.

The minority Liberal government that assumed office in 1985 with the support of the social democratic NDP represented a significant break with the previous forty years of Conservative rule. One of the first steps taken by the Liberal Premier was to announce the formation of the Premier's Council in April, 1986, a multipartite body charged with responsibility for analysing the competitive strengths of the provincial economy, determining its weaknesses, and proposing new policy directions for Ontario. Two years later, the Council produced a path-breaking study entitled *Competing in the New Global Economy*. The Premier's Council Report released in April 1988 presented the most coherent vision of an industrial and technology strategy espoused by either of the two mainstream parties in the province. The study was noteworthy for the dramatic break it constituted with the prevailing neo-conservative orthodoxy in North America. Rather than viewing international competition as a threat to workers' wages, the Premier's Council depicted effective international competition as the key to a high wage economic strategy and higher standards of living,

> In the future our prosperity will depend increasingly upon our ability to sustain a sufficiently large base of companies competing in world markets, not on the basis of lower labour or raw materials costs, but rather through technological innovation, skilled labour, adept marketing, and high productivity . . .
>
> (Premier's Council 1988, 35–7)

The Premier's Council counterposed to this a strategy of achieving higher value-added per employee by realizing higher productivity in the existing mix of goods and services, or by shifting the industrial structure of the province towards goods and services that are inherently higher value-added per employee activity. The Council argued that this strategy could best be pursued by focusing its economic policies on traded businesses – those exposed to world trade and competition. Gains realized through improvements in traded goods and services would generate increased prosperity throughout the provincial economy. Hence, 'the traded sectors must be viewed as the fundamental drivers of our future wealth and prosperity' (1988: 41). The key to accomplishing this goal was to increase the number of indigenous world-scale companies capable of competing effectively in global markets. The Council viewed indigenous firms as more likely to provide higher value-added jobs, generate indirect employment, and create spin-off companies in the province. The Council defined Ontario's challenge as needing to accelerate the growth of indigenous threshold firms in the traded sectors with the potential to reach world-scale levels of activity, while expanding the strengths of the existing non-indigenous base of companies (1988: 75).

Not content to wait for the results of the Council's deliberations, however, the Liberal government simultaneously announced the creation of a ten-year $1 billion Technology Fund in 1986 to be partly under the direction of the Premier's Council. The speedy implementation of a wide range of programs under the aegis of the Technology Fund stands as one of the lasting achievements of the government. The first program to be revamped and brought under the umbrella of the Technology Fund was the University Research Incentive Fund (URIF) initiated in 1984. URIF was designed to encourage universities to collaborate with the private sector in joint research initiatives. It provided matching funding for private sector investments in short-term university contract research. The level of funding varied from year to year, but averaged between $5 and $6 million. In June 1987 the Council designated seven provincial Centres of Excellence to carry out long-term basic research in Ontario universities. Each of the seven designated centres involves collaborative research between faculty at more than one university and industry partners. Funded at an initial level of $204 million over five years, the centres were as follows: Advanced

Laser and Lightwave Research, Space and Terrestrial Science, Integrated Manufacturing, Groundwater Research, Information Technology, Materials Research and Telecommunications Research. The third major component announced by the government in December 1987 was the Industry Research Program (IRP). The IRP was designed to stimulate leading edge industrial R&D in areas of strategic importance with a demonstrated export or import replacement potential. Successful applicants were required to fund a portion of the project as well. At its peak in the early 1990s, the IRP was providing funding for twenty-four separate projects with a total annual expenditure of $23 million (Premier's Council on Economic Renewal, 1993).

TOWARDS INSTITUTIONAL REFLEXIVITY

The election of the New Democratic Party government in Ontario in September 1990 was regarded as one of the great political upsets in the province's history. Few observers, even those closest to the party's campaign, seriously believed it would win until a few short days before the election. The NDP took office just as the worst recession since the 1930s hit the province. While some had foreseen that the Free Trade Agreement with the US would force a major adjustment upon the provincial economy, few anticipated that it would occur in the context of high real interest rates, an overvalued Canadian dollar, and a severe cyclical recession. The recovery which began in 1991 was hesitant at best and the overall level of employment did not surpass the 1989 peak until July 1994.[5]

The NDP, both federally and provincially, had long advocated the adoption of an industrial strategy, but was somewhat vague on specifics. Policy resolutions adopted at party conventions regularly called for the nationalization of leading corporations, but there was little active consideration of this option by the ministers of the new government. Those political advisers and senior bureaucrats responsible for formulating economic policy were determined to build on the foundations of a more active industrial and technology policy begun by their predecessors, but to avoid some of the problems encountered in the past. The literature on the potential value of regional networking and cooperation in enhancing competitiveness provided a useful starting point to build that alternative (Best, 1990). The development of a regional perspective on industrial policy in Ontario thus entered its third phase.

One of the first opportunities to articulate an alternative economic vision came with the government's first budget of April 1991. Although the budget itself was widely condemned in the press for its naïve Keynesianism, a background paper outlined a somewhat more sophisticated approach to economic policy. The budget paper, *Ontario in the 1990s*, established sustainable prosperity based on equitable structural change as its goal. The paper recognized that competitiveness was important to attaining this objective, but argued that it could best be realized through the creation of high value-added, high wage jobs and creating strategic partnerships. The key to long-term competitiveness, according to the budget paper, involved a number of factors, including the ability to improve productivity performance; the skills and adaptability of the labour force; the quality of management skills; the capacity for innovation, institutional flexibility and a strong foundation of physical and social infrastructure. However, at the root of all these factors was a recognition of the increased importance of knowledge-based inputs, or technology, in every aspect of the production process. It acknowledged that technological change is a social process integrally tied to every phase of economic activity. As such, the challenge for public policy was to promote an organizational culture supportive of technology and innovation in the workplace and society at large (Laughren, 1991).

Some of the themes introduced in this budget paper were reinforced and expanded upon in the succeeding budget of 1992. The supplementary budget paper, 'Investing in Tomorrow's Jobs', argued that

> Investment in plant, technology, training, infrastructure and flexible organizations is critical for renewed productivity growth in Ontario. This requires a joint effort among business, labour and government. Government programs and private activities must be redirected to promote the kinds of change that lead to a new, more flexible and knowledge-intensive economy.
>
> (Laughren 1992, 19)

The government adopted the phrase economic renewal to refer to those policies and programs with the potential to encourage the long-term restructuring of the provincial economy towards the creation of more well-paid, high value-added jobs.

At the centre of the economic renewal agenda was the government's Industrial Policy Framework, released in July 1992 by the Minister of Industry, Trade and Technology. The Industrial Policy Framework built on the analysis of the Premier's Council report, but supplemented it with insights drawn from the experience of some US states and the more innovative subnational jurisdictions in Europe. The overall goal of the framework was to promote the transition of the Ontario economy towards those sectors and firms with the capacity to generate higher wage, higher value-added and environmentally sustainable jobs. The framework focused on ways of developing higher value-added activities throughout the economy to increase competitiveness and create more, and better, jobs. It consisted of three main elements:

1 changing the way government invests, including measures to enhance the quality of physical and technological infrastructure through institutions such as the seven provincial Centres of Excellence (funding for the Centres was renewed in July 1992 for an additional five years);
2 changing the way government works with companies, specifically through the negotiation of strategies developed in cooperation with the firms and unions in specific sectors to improve their competitiveness. This goal is also to be achieved through focusing the government's support on strategic groups of firms, such as those committed to continuous innovation and to improving their technological capabilities; and,
3 changing how government responds to economic change, through measures to support investment in green industries and enhanced support for organizational change in Ontario firms (Ministry of Industry 1992).

The most significant change envisioned in the framework was the increased emphasis placed on working with sectors. The focus on sectors was not an entirely novel departure. The Ministry of Industry, Trade and Technology (MITT) had a tradition dating back to the early 1980s of working with individual sectors to strengthen their competitive position and promote their sectoral capabilities. Notable examples of this activity included efforts to strengthen the position of Ontario's film and television industry, the radical restructuring of the wine industry under the Ontario Winery Adjustment Program and the food processing development strategy previously recommended by the Food Industry Advisory Committee. In addition, MITT had developed an internal sectoral capacity through its dealing with individual sectors, such as the automotive industry and the furniture industry. Finally, the Liberal government had also established an Industrial Restructuring Commissioner whose responsibilities included the adjustment problems faced by specific sectors.

An early example of the kind of sectoral initiatives contemplated by the Industrial Policy Framework was the Memorandum of Understanding signed by the Minister of Industry with the Society of the Plastics Industry of Canada (SPI) in 1991. The SPI is one of the best organized and most politically astute industry associations in the country. It had worked previously with the federal and Quebec governments to develop policies designed to support the sector. Its early desire to enter into the Memorandum of Understanding with Ontario anticipated the government's emerging emphasis on sectors in the Industrial Policy Framework. The Memorandum confirmed the mutual objectives of the SPI and the government of Ontario to work together in partnership to make the Ontario plastics industry internationally competitive. To that end, the government agreed to support three individual projects endorsed by the SPI: the establishment of the Industrial Research and Development Institute in Midland Ontario, a national institute for applied R&D and technology diffusion for the tool, die, mould, and related manufacturing and material industries; the construction of a new Technology Centre by Husky Injection Molding in Bolton, Ontario which will benefit a wide range of firms in the industry; and the establishment of the Canadian Plastics Training Centre at Humber College in Toronto.

The Sector Partnership Fund announced in the budget of April 1992 was a three-year initiative, budgeted at $150 million, and designed to implement the sectoral component of the Industrial Policy Framework. The Sector Partnership Fund provided assistance to approved cooperative sector projects leading to higher value-added activities. It recognized that each sector faces unique competitive challenges and was designed to respond to those circumstances. For the purposes of the sector development process, a sector was defined as a group of Ontario-based firms that produce similar goods and services and that identified themselves as a sector, have a recognized association or forum for resolving sector-specific issues, have identified a range of sectoral issues of concern to a broad cross-section of members, and have multipartite representation, including business, labour and other relevant stakeholders in the sector.

The Sector Partnership Fund (SPF) was based on the four principles of flexibility, cooperation, leverage and accessibility. It recognized that each sector faces unique competitive challenges and was designed to be flexible in its response to those circumstances. Individual industrial sectors are characterized by distinctive sectoral properties, shaped by the specific nature of the technology they use and the constraining effects of their products and product markets. A critical principle incorporated into the SPF was that of leverage. In a time of scarce fiscal resources, the government maintained that it could not, and should not, assume full responsibility for funding sector-based initiatives. SPF support was intended to lever project funding from industry, labour and other levels of government. The process was based, in part, on the assumption that eligible projects constituted a form of quasi-public goods, whose utility to industry partners was strong enough to attract some private investment, but insufficient to be self-financing. It was also seen as a way of subjecting the sectoral initiatives to a form of market test to determine if the private sector was willing to support them itself. The third principle followed Michael Best's argument in assuming that within individual sectors, competition and cooperation should be viewed as complementary activities. Finally, the principle of accessibility established that all sectors were deemed potentially eligible for funding and that within each sector a substantial proportion of its participants must stand to benefit from SPF-supported initiatives.

While the hallmark of the SPF was to choose winning activities, not winning sectors, the government did designate a number of areas as priority sectors for policy development. One of these was Green or Environmental Industries. Green industry involves products and processes that contribute to energy efficiency, water conservation, pollution prevention and the three R's. The thrust of the strategy was to foster the development of green products by Ontario industries, as well as to create a global demand for these products. The goal was to support Ontario firms in manufacturing the green products and intermediate goods that will be stimulated by a leadership approach on environmental standards and energy efficiency. While this strategy was slow to evolve, early measures focused on using energy efficiency and environmental retrofit programs as a way of leveraging demand for green products within the province. In addition, several sector strategies initiated by the previous government were folded into the framework of the SPF. One of these was the Food Processing Industry, where the Liberal government had approved a sector strategy developed by an advisory committee (The Food Industry Advisory Committee, 1990), and a food industry competitiveness branch had been created in the Ministry of Agriculture and Food. The Liberal government had also approved a competitiveness strategy for the grape and wine industry in 1989 and provided a $45 million winery assistance program to support the industry's efforts to move towards higher quality, more internationally competitive products. In each case, initiatives brought forward under these respective strategies have qualified for funding from the SPF.

By any criteria of measurement, the initial stage of sector consultation and strategy formation must be viewed as a success. Both the number of sectors involved and the extent of participation by key sector players in the consultative process exceeded the expectations of government officials by a wide margin. Between the summer of 1992 and the provincial election in June 1995, the Ministry of Economic Development and Trade, along with the other lead ministries with sector responsibility, worked with a wide range of industry associations, trade unions and other stakeholders to develop sector strategies. Consultative efforts produced approved strategies in fifteen sectors: Food Processing, Green Industries, Telecommunications, Computing, Tourism, Cultural Industries, Aerospace, Autoparts, Mines and Minerals, Construction, Health Industries, Forestry, Plastics, Residential Furniture and Chemicals. By the spring of 1995, work plans were approved and strategies under development in a range of additional sectors, including: Biotechnology, Consulting Engineering, Design, Machinery, Tool, Die and Mould, Retail, and the Electrical and Electronics industry. The last of these strategies was released formally in May 1996.

In each case, the consultative efforts were broad and inclusive, drawing in as many as 150 individuals in the sector to prepare detailed analyses of sectoral strengths and weaknesses and propose a course of action. In total, the approach involved twenty-eight different sectors and over two thousand individual participants. In the process, the sector partners included twenty-two different unions, ninety-three industry associations and twenty-eight universities and colleges (Ontario, 1995). This was a major accomplishment for the process, given the continuing opposition by the business community to most of the government's policy agenda.

One notable example of this success was the report of the Advisory Committee on a Telecommunications Strategy for Ontario presented to the Minister of Culture and Communications in August 1992. This report set out a bold and challenging vision for the telecommunications industry of the province. The government responded to its recommendations by establishing an Ontario Network Infrastructure Program (ONIP) in February 1993 to accelerate the development of telecommunications-based information networks in the province. It also announced the establishment of four priority areas for funding under the SPF: developing new telecom applications; forming innovative business enterprises; establishing specialized sectoral infrastructure; and enhancing market development and promotion. A number of initial projects have been funded under this program, including OCRInet, a wide area network linking a number of post-secondary research institutions and private research centres in the Ottawa–Carleton region and several local community-based information networks.

The high number of sectors that participated in the strategy development process and their relative success in achieving consensus on their strategic plans would suggest that demands on the Sector Partnership Fund should have been high. Indeed there was no lack of recommendations for concrete initiatives in virtually all of the plans. These initiatives tended to be grouped into four categories: access to capital, technology and R&D, education and training and export trade development. Despite this fact, the Sector Partnership Fund underspent its allocation in every year that it existed, and at the time of its termination in July 1995 little more than half of the $150 million allocation had been committed. A number of factors accounted for this outcome. One that created a substantial barrier was the expectation of industry funding for the initiatives. The imposition of the 'quasi-market test' on SPF initiatives clearly imposed a hurdle that many private sector participants had difficulty surmounting.

Despite these problems, a number of significant initiatives did receive funding from the SPF. Many of the initiatives that received funding were oriented towards the creation of sector-based technology centres. Examples included the Guelph Food Technology Centre, designed to increase effective technology and information transfer, as well as to provide accessible pilot plant facilities for the food industry; an Ontario Centre for Environmental Technology Advancement to provide technical support services, financial advice and business counselling to help young firms commercialize environmental technologies; Connect-IT, a computing Sector

Resource Facility to assist the many small and medium-sized firms in Ontario's industry in developing sector-specific competency in management, standards, marketing expertise and export readiness. In the computing sector funding was also provided to support the Electronic Commerce Institute to promote the adoption and use of electronic data interchange in Canadian industry. Other areas that received some funding included export market development through Interhealth Canada, a private, not-for-profit corporation designed to pursue and gain international contracts for Canadian firms in key markets around the world; and the plan to establish representatives for the autoparts sector in Japan and Europe to help increase sales to Japanese and European assemblers in their North American and foreign operations.

In addition to its sectorally related activities, the government also took measures to assist smaller, innovative high technology companies. One such measure was the creation of the Ontario Innovation and Productivity Service (OIPS). Modelled after aspects of several other institutions, including the Michigan Modernization Service, the Australian National Industrial Extension and elements of Ontario's own Manufacturing Recovery Program, OIPS was designed to help innovative growth firms overcome barriers to their further expansion. OIPS worked with a selective group of target firms to help them identify challenges and opportunities for growth and develop a strategic business plan. It provided funding for strategic projects and facilitated enhanced access to other provincial government programs, as well as to brokers with the federal government and existing private sector resources. During its brief existence, the program assisted over 300 small and medium-sized firms with innovation in the areas of product development, design, marketing and production at an average cost of $65,000.

Under both the Liberal and NDP governments, a growing number of sector-based institutions in the areas of industrial policy, labour market policy and financing mechanisms were created, raising some interesting issues in terms of coordinating the prospective elements of the regional innovation system. A report prepared on the economic impact of the Centres of Excellence (COEs) drew attention to the desirability of viewing these discrete initiatives in the area of industrial and technology policy as parts of a complex and integrated provincial or regional system of innovation. The report emphasized the need to develop criteria for measuring the effectiveness of the COEs as linkages between other elements of the regional innovation system, particularly the science and technology infrastructure, the educational system, business enterprises and government (The Impact Group, *et al.*, 1994). While this analysis was limited to the COEs, there is a strong case to be made for using a similar framework to assess the degree of integration among the other elements of the system and target areas in need of more effective coordination. This is clearly a prerequisite for Ontario to become an *intelligent region* displaying the kind of *institutional reflexivity* espoused by Cooke.

The considerable progress made in this direction during the Liberal and NDP interregnum came to an abrupt end with the election of a new Conservative government in 1995. The Tory platform, labelled the Common Sense Revolution (CSR), called for an abrupt shift in the direction of government spending in general, and its economic development policies in particular. The CSR displayed a strong preference for the use of broad framework policies, such as a reduction in the tax and regulatory burden to stimulate growth, in contrast to the more targeted spending policies favoured by the Premier's Council and the NDP government. Within the first six months of assuming office, nearly all of the initiatives put in place by the two previous governments were cancelled or wound down. Virtually all that remained were the Centres of Excellence, and their survival was not assured until December 1996 when the Minister of Economic Development announced their renewal for the next five years with an annual budget of $32.5 million and a consolidation of the seven centres down to four. The Minister indicated that the renewed COEs would be tied more closely to the private sector for the purpose of promoting economic growth and job creation. Their primary purpose was to encourage collaboration between universities and the private sector through partnerships that give industry better access to university research expertise.

For the first two years of its mandate the issue of industrial and technology policies took a back seat, as the Conservative government struggled with a broader agenda of reducing expenditures and personal income taxes, reforming the welfare system and engineering a massive redistribution of responsibility between the provincial and local governments. It was not until the 1997 budget that some of the elements of the province's industrial and technology policy re-emerged on the agenda. The budget contained a major new spending program and several additional tax incentives. In addition, it included a long budget paper on the benefits of investing in innovation and R&D for the Ontario economy and a rationale for the new spending program and tax incentives being introduced with the budget. Both the budget speech and the accompanying budget paper were filled with references to the need for increased partnership between the private sector and other elements of the innovation system, especially the university-based research institutions.

Of the changes introduced, the CFI's $500 million R&D Challenge Fund is the most significant. Its primary rationale is to promote business–university partnerships and research excellence. The stated objectives of the fund are to support job creation and economic growth; to promote world-class research of interest to the private sector; and to encourage more collaboration between the private sector and research institutions. According to the government, the main priority of the fund will be to attract and keep world-class researchers in Ontario. It will have the flexibility to provide support for leading edge research that benefits today's growing industries and helps create the industries of the future; state-of-the-art equipment and facilities; and incentives for gifted researchers to work in Ontario, including endowed chairs. The funding will be awarded on a competitive basis, according to the proposal's contribution to research excellence and economic growth. One criterion of its economic benefit will be the ability to attract private sector support. However, the fund will also be used to allow provincial research centres to match funds from other federal and international granting councils. Another of its stated objectives is to ensure that Ontario's universities are able to compete for funding from the federal CFI on a timely basis. Of the various tax incentives introduced in the budget, the most significant is the Ontario Business-Research Institute Tax Credit – a 20 per cent refundable R&D tax credit for corporate-sponsored R&D performed in Ontario by eligible universities or other approved post-secondary educational institutes or research associations (Eves, 1997, 177–83). While these measures do not offset the full weight of the programs and incentives eliminated in the first year of the government's mandate, they do indicate a more substantial interest in a cooperative approach to industrial and technology policy than was the case in the first two years of the government's life.

CONCLUSION

This brief survey of industrial and technology policies for both the federal government and Ontario reveals a pattern of important initiatives, undermined by inconsistency and continuous policy shifts. At both levels of government, there have been significant developments in the past decade directed at strengthening the role of public sector institutions and policies in the national and regional systems of innovation. Many of the initiatives have been directed towards strengthening individual elements of the system and improving the overall cohesion and integration of both the public and private sector components. As such there has been a strong emphasis on criteria such as increasing cooperative research efforts among private sector firms, working with sectors, promoting the linkages between the public research organizations and private sector firms and improving the access to, and involvement of, private firms with the university research base. Despite the considerable progress made in this direction over the past decade, many of the positive steps taken have been undermined by frequent shifts in policy direction and rapid changes in funding levels both between governments of differing party composition and, on occasion, within the same government.

As a consequence, it would be an overstatement to suggest that the policies implemented at the federal level or in Ontario qualify either jurisdiction for inclusion in the select group of national or regional states advancing the frontier of innovative industrial and technology policy. Yet the initiatives taken over the past decade describe a somewhat halting and inconsistent process of moving towards a more reflexive approach to their respective systems of innovation. Over this period, both levels of government and some industry associations have recognized the need for a more coordinated and integrated approach to developing and applying new knowledge and ideas within the system of innovation. More substantial progress has been undermined by the overriding preoccupation with fiscal problems and a lack of consensus between the various parties over the most effective approach to adopt. However, they have also been hampered by the lack of coordination, and, even, by duplication of efforts between the two levels of government. A more effective approach will require greater policy consistency over a longer period and better efforts at coordination between the federal and provincial governments.

The policy developments outlined above suggest that Ontario, and some of the other provinces as well, have begun to adapt their industrial and technology policies in light of the lessons learned from other regions abroad. While these initiatives may fall short of the full requirements of *institutional reflexivity*, they point the way in a more active direction. The way towards a truly cooperative approach to industrial policy in Canada requires a greater recognition of this shift on the part of the federal government and a more consistent effort to integrate both its own efforts and those of the provinces into a regionally based, but national system of innovation. As Michael Jenkin observed a decade ago,

> Perhaps the best way to deal with the more active industrial policy role played by the provinces is to view provincial industrial strategies as an opportunity rather than a liability in the creation of a stronger and more aggressive response to restructuring the national economy.
>
> (Jenkin, 1983, 172)

NOTES

1 A third approach to the concept can be found in the work of Richard Nelson who tends to see national systems of innovation in more institutional terms, looking at the role of government policy at the national level, formal regulation and informal coordination, levels of R&D funding and the supporting role played by the educational infrastructure (Nelson, 1988, 1993). For a recent overview and synthesis of the different approaches to the systems of innovation concept, cf. Edquist (1997).

2 Ironically, at least one major firm in the information technology sector reported in an interview that the tightness of criteria around networking and strategic alliances made it very difficult to qualify for funding and accounted for the slow pace of take-up in this stream of the program.
3 The preceding argument draws in part on my recent article Wolfe (1997).
4 For a preliminary overview of a range of the relevant policies, cf. Wolfe (1994).
5 For a fuller discussion of the economic and fiscal problems encountered by Ontario's first NDP government, cf. Rachlis and Wolfe (1997).

BIBLIOGRAPHY

Arthur, W. Brian, *Increasing Returns and Path Dependence in the Economy*, 'Foreword' by Kenneth Arrow (Ann Arbor: The University of Michigan Press, 1994).

Bell, Daniel, *The Coming of Post-industrial Society: A Venture in Social Forecasting* (New York: Basic Books, 1973).

Best, Michael H., *The New Competition: Institutions of Industrial Restructuring* (Cambridge: Polity Press, 1990).

Botham, Richard and Giguère, Michel, 'The Industrial Research Infrastructure', in P. Dufour and J. de la Mothe, eds, *Science and Technology in Canada* (Harlow: Longman, 1993).

Cooke, Philip, 'Institutional Reflexivity and the Rise of the Region State', in G. Benko and U. Strohmayer, eds, *Space and Social Theory: Interpreting Modernity and Post-modernity* (Oxford: Blackwell, 1997).

Cooke, Philip, 'Introduction: Origins of the Concept', in H.-J. Braczyk, P. Cooke and M. Heidenreich, eds, *Regional Innovation Systems: The Role of Governances in a Globalized World* (London: UCL Press, 1998b).

Cooke, P. and Morgan, K., 'The Network Paradigm: New Departures in Corporate and Regional Development', *Environment and Planning D: Society and Space*, Volume 11: 543–64, 1993.

de la Mothe, John, 'One Small Step in an Uncertain Direction: The Science and Technology Review and Public Administration in Canada', *Canadian Public Administration*, Volume 39: 403–17, Autumn 1996.

Doern, G., Bruce, P. and Tomlin, Brian W., 'Trade–Industrial Policy', in G. Doern, P. Bruce, A. Leslie and Brian W. Tomlin, eds, *Border Crossings: The Internationalization of Canadian Public Policy* (Toronto: Oxford University Press, 1996).

Dosi, Giovanni, 'Sources, Procedures and Microeconomic Effects of Innovation', *Journal of Economic Literature*, Vol. XXVI: 1120–71, September 1988.

Dosi, Giovanni, Freeman, Christopher, Nelson, Richard, Silverberg, Gerald and Soete, Luc, eds, *Technical Change and Economic Theory* (London and New York: Pinter Publishers, 1988).

Drucker, Peter F., *Post-capitalist Society* (New York: HarperBusiness, 1993).

Dufour, Paul and de la Mothe, John, eds, *Science and Technology in Canada* (Harlow, Longman, 1993a).

Dufour, Paul and de la Mothe, John, 'The Organisation, Structure and Decision Processes of Science and Technology', in P. Dufour and J. de la Mothe, eds, *Science and Technology in Canada* (Harlow: Longman, 1993b).

Edquist, Charles, 'Introduction: Systems of Innovation Approaches – Their Emergence and Characteristics', in Charles Edquist, ed., *Systems of Innovation: Technologies, Institutions and Organizations* (London: Pinter Publishers, 1997).

Eves, Ernie, 'The R&D Opportunity: Cutting Taxes and Creating Jobs', *1997 Ontario Budget*, Budget Paper E. (Toronto: Queen's Printer for Ontario, 6 May 1997).

Florida, Richard, 'Toward the Learning Region', *Futures*, Vol. 27: 527–36, May/June 1995.

The Food Industry Advisory Committee, *Report* (Toronto: Queen's Printer for Ontario, 1990).

Freeman, Chris, 'The "National System of Innovation" in Historical Perspective', *Cambridge Journal of Economics*, Vol. 19: 5–24, 1995.

Freeman, Christopher, *Technology Policy and Economic Performance: Lessons from Japan* (London and New York: Pinter Publishers, 1987).

Freeman, Christopher and Perez, Carlota, 'Structural Crises of Adjustment, Business Cycles and Investment Behaviour', in Giovanni Dosi, Christopher Freeman, Richard Nelson, Gerald Silverberg and Luc Soete, eds, *Technical Change and Economic Theory* (London and New York: Pinter Publishers, 1988).

Gibson, David V. and Rogers, Everett M., *R&D Collaboration on Trial: The Microelectronics and Computer Corporation* (Boston: Harvard Business School Press, 1994).

Government of Canada, *Science and Technology for the New Century: A Federal Strategy* (Ottawa: Supply and Services Canada, 1996).

Government of Canada, Industry Canada, *Building a More Innovative Economy* (Ottawa: Supply and Services Canada, 1994).

Grindley, Peter, Mowery, David C. and Silverman, Brian, 'The Design of High-Technology Consortia: Lessons from SEMATECH', in Morris Teubal, Domique Foray, Moshie Justman and Ehud Zuscovitch, eds, *Technological Infrastructure Policy: An International Perspective* (Dordrecht/Boston/London: Kluwer Academic Publishers, 1996).

Grossman, Hon. Larry, *Economic Transformation: Technological Innovation and Diffusion in Ontario* (Toronto: Queen's Printer for Ontario, 1984).

The Impact Group, Hickling, KlasTek Ltd and IFIAS, *Innovation and Wealth Creation: Performance Measures for the Ontario Centres of Excellence* (Toronto, 1994).

Jenkin, Michael, *The Challenge of Diversity: Industrial Policy in the Canadian Federation*, Science Council of Canada Background Study 50 (Ottawa: Supply and Services Canada, 1983).

Laughren, Floyd, 'Ontario in the 1990s: Promoting Equitable Structural Change', *1991 Ontario Budget*, Budget Paper E (Toronto: Queen's Printer for Ontario, 29 April 1991).

Laughren, Floyd, 'Investing in Tomorrow's Jobs: Effective Investment and Economic Renewal', *1992 Ontario Budget*, Supplementary Paper (Toronto: Queen's Printer for Ontario, 30 April 1992).

Lundvall, Bengt-Åke, 'Innovation as an Interactive Process: Form User–Producer Interaction to the National System of Innovation', in Giovanni Dosi, Christopher Freeman, Richard Nelson, Gerald Silverberg and Luc Soete, eds, *Technical Change and Economic Theory* (London and New York: Pinter Publishers, 1988).

Lundvall, Bengt-Åke, 'Introduction', in Bengt-Åke Lundvall, ed., *National Systems of Innovation: Towards a Theory of Innovation and Interactive Learning* (London: Pinter Publishers, 1992a).

Lundvall, Bengt-Åke, ed., *National Systems of Innovation: Towards a Theory of Innovation and Interractive Learning* (London: Pinter Publishers, 1992b).

Lyrette, Jacques, 'The Synergy of Industry with Canadian Research Institutions', Unpublished draft, n.d.

McFetridge, Donald G., 'The Canadian Industrial System of Innovation', in Richard R. Nelson, ed., *National Innovation Systems: A Comparative Analysis* (New York and Oxford: Oxford University Press, 1993).

Miller, Hon. Frank S., *R&D and Economic Development in Ontario: A Discussion Paper* (1983 Ontario Budget Paper) (Toronto: Queen's Printer for Ontario, 1983).

Ministry of Industry, Trade and Technology, *An Industrial Policy Framework for Ontario* (Toronto: Queen's Printer for Ontario, 1992).

Murray, Robin, *Local Space: Europe and the New Regionalism*. (Manchester and Stevenage: Centre for Local Economic Strategies and South-East Economic Development Strategy, 1991).

Mytelka, Lynn Krieger, 'Knowledge-Intensive Production and the Changing Internationalization Strategies of Multinational Firms', in James A. Caporaso, ed., *A Changing International Division of Labour* (Boulder, Colorado: Lynne Reiner, 1987).

Nelson, Richard R., 'Institutions Supporting Technical Change in the United States', in Giovanni Dosi, Christopher Freeman, Richard Nelson, Gerald Silverberg and Luc Soete, eds, *Technical Change and Economic Theory* (London and New York: Pinter Publishers, 1987).

Nelson, Richard R., ed., *National Innovation Systems: A Comparative Analysis* (New York and Oxford: Oxford University Press, 1993).

Niosi, Jorge, Saviotti, Paolo, Bellon, Bertrand and Crow, Michael, 'National Systems of Innovation: In Search of a Workable Concept', *Technology in Society*, Vol. 15: 207–27, 1993.

Ohmae, Kenichi, 'The Rise of the Region State', *Foreign Affairs*, Vol. 72: 78–87, 1993.

Ontario, Ministry of Economic Development and Trade, *Ontario Sector Snapshots: A Progress Report on the Sector Development Approach* (Toronto, 1995).

Organization for Economic Co-operation and Development, *New Technologies in the 1990s: A Socio-economic Strategy* (Report of a Group of Experts on the Social Aspects of New Technologies) (Paris: OECD, 1988).

Paquet, Gilles and Roy, Jeffrey Roy, 'Prosperity Through Networks: The Bottom-up Strategy That Might Have Been', in Susan D. Phillips, ed., *How Ottawa Spends 1995–96: Mid-life Crises* (Ottawa: Carleton University Press, 1995).

Premier's Council, *Competing in the New Global Economy* (Vol. 1) (Toronto: Queen's Printer for Ontario, 1988).

Premier's Council on Economic Renewal, *Ontario 2002: A Report of the Task Force to Review the Ontario Technology Fund in the Context of an Innovation-based Society* (Toronto: Queen's Printer for Ontario, 1993).

Putnam, Robert D. (with Robert Leonardi and Rafaella Y. Nanetti) *Making Democracy Work: Civic Traditions in Modern Italy* (Princeton, NJ: Princeton University Press, 1993).

Rachlis, Chuck and Wolfe, David, 'An Insider's View of the NDP Government of Ontario: The Politics of Permanent Opposition Meets the Economics of Permanent Recession', in Graham White, ed., *The Government and Politics of Ontario* (5th edn) (Toronto: University of Toronto Press, 1997).

Rea, K. J., *The Prosperous Years: The Economic History of Ontario, 1939–1975*, The Ontario Historical Studies Series (Toronto: University of Toronto Press, 1985).

Sabel, Charles F., 'Flexible Specialisation and the Re-emergence of Regional Economies', in Paul Hirst and Jonathan Zetlin, eds, *Reversing Industrial Decline* (Oxford: Berg, 1989).

Scott, Allen J., 'The Roepke Lecture in Economic Geography – The Collective Order of Flexible Production Agglomerations: Lessons for Local Economic Development Policy and Strategic Choice', *Economic Geography*, Vol. 68, No. 3; 219–33, July 1992.

Scott, Allen J. and Storper, Michael, 'Regional Development Reconsidered', in Huib Ernste and Verena Meier, eds, *Regional Development and Contemporary Industrial Response: Extending Flexible Specialisation* (London and New York: Belhaven Press, 1992).

Storper, Michael, 'The Resurgence of Regional Economies, Ten Years Later', *European Urban and Regional Studies*, Vol. 2, No. 3: 191–221, 1995.

Tapscott, Don, *The Digital Economy: Promise and Peril in the Age of Networked Intelligence* (New York: McGraw-Hill, 1996).

Wolfe, David A., 'The Wealth of Regions: Rethinking Industrial Policy', Canadian Institute for Advanced Research, Program in Law and the Determinants of Social Ordering, *Working Paper No. 10* (Toronto, Ontario, May 1994).

Wolfe, David A., 'The Emergence of the Region State', in Thomas J. Courchene, ed., *The Nation State in a Global/Information Era: Policy Challenges*, The Bell Canada Papers on Economic and Public Policy 5 (Kingston: John Deutsch Institute for the Study of Economic Policy, Queen's University, 1997).

3

COMMENTARY

Michael J. Trebilcock

Robert Howse's chapter makes a compelling case, in the Canadian context, against a major emphasis on targeted public support of high technology industries – in part because it is not clear that there are major market failures that private economic agents cannot solve for themselves, and in part because even where there is some evidence of market failure, the political economy of government intervention in a decentralized federal state such as Canada is likely to make it extremely difficult to implement and sustain a set of coherent industrial policies directed to promoting high technology industries. Hence, my comments on the chapter are of a second-order nature.

The claim made for activist government industrial policies with respect to David Crane has also been made by other commentators. I am thinking of Richard Harris's study for the Macdonald Commission, various publications of the now defunct Science Council of Canada, Dr. Fraser Mustard of the Canadian Institute of Advanced Research, and scholars associated with CIAR projects (e.g. Richard Lipsey). The chapter would be strengthened by a broader set of references to the views of these various parties.

The positive externalities argument for government intervention in high-tech sectors examined by Howse is largely identified with Michael Porter's theory of economic clusters. There is, however, a broader literature on Modern (or Endogenous) Growth Theory, which emphasizes the role of technological innovation and application in promoting economic growth and which is open to an activist policy for the state in realizing the dynamic economics associated with the introduction and exploitation of modern technology and its associated externalities and spillovers.[1]

I think Howse perhaps needs to emphasize more sharply (than he does) that there are two conceivable roles that government policy might play in promoting cooperation in industrial policy. First, simply, is to get out of the way, for example, by appropriately relaxed application of anti-trust laws to cooperative research arrangements. While Howse mentions this possibility, there is indeed a good deal of controversy in the economic literature on whether cooperative research and production arrangements are, or are not, welfare-enhancing. The Director of Investigation and Research of the Canadian Competition Policy Bureau has recently released a set of Guidelines on Strategic Alliances that indicates his current thinking on the application of various provisions in the Competition Act to these kinds of arrangements. The second role of government, on which Howse focuses principally, is a proactive role, where he rightly argues that targeted forms of state assistance to high-tech sectors are likely to have highly ambiguous, and often negative, welfare effects. With respect to a more proactive role, it would be useful for him to mention the various international trade policy issues that are likely to arise here. For example, if government is actively involved in promoting various kinds of cooperative research arrangements, there may well be issues of discrimination that arise under the new Uruguay Round Agreement if foreign firms are excluded from participation in these arrangements. In addition, if subsidies are involved, particularly if the industries in question are export-oriented, then the very tight new strictures on export subsidies under the Uruguay Round Subsidies Agreement may make such a strategy more vulnerable to challenge than has been the case in the past.

Howse makes the point that total Canadian spending on R&D (measured as a per centage of GDP) has lagged behind that of many other industrialized countries. While this is true, it is important to point out that government spending on R&D in Canada constitutes a high percentage of total spending on R&D, and compares favourably with most other industrialized countries. The low level of spending on R&D is largely attributable to low levels of spending by the private sector. Whether these low levels of private sector investment are socially suboptimal is not clear. Nor is it at all clear why government should increase its contribution above the high levels that already prevail.

Howse makes the point that it is not true that in the past Canadian industrial policy has been about protecting losers rather than picking winners. He cites several examples of government attempts to pick winners. There are a number of other examples that might be cited (e.g. Deuterium and Clairtone), a number of which I and several colleagues reviewed in a study of bailouts in Canada completed in the early 1980s for the now defunct Ontario Economic Council. It is of course true that, in addition to failed efforts to pick winners, a large part of Canadian industrial policy in the past has been about protecting declining sectors, where the record is at least as bleak.

Howse refers to the important role that government can play in supporting non firm-specific education and training, and in that context mentions briefly the possible virtues of federal funding and training through vouchers rather than transfers to the provinces or to provincial institutions. Here it is worth pointing out that Lloyd Axworthy, then Minister of Human Social Resources, proposed a similar idea in a policy paper in

1995, which was greeted with a storm of criticisms from various constituencies, and seems to have been abandoned. Some exploration of the nature of these objections may be worthwhile.

Howse refers to the importance of civic institutions, what Putnam calls 'social capital' in his celebrated book, *Making Democracy Work: Civil Traditions in Modern Italy*. Here, I remain to be persuaded that public provision or underwriting of child care has much to do with social capital; nor is there overwhelming evidence that child care is grossly underprovided at present, or that governments in Canada can afford to embark upon a massive new social program. On the other hand, like Howse, I see virtues in governments exploring more vigorously the potential for delivering social and adjustment programs through community-based groups and organizations, although it is not clear that this is the principal function of social capital emphasized by Putnam, who sees civic organizations more as a mechanism for ensuring accountability of government for the performance of its functions.

A more general issue that Howse does not explicitly address in his chapter, which I think is of crucial importance, is the definition of what constitutes a high technology sector. Debates over promoting high technology sectors often proceed as if one can simply assume the definitional ambit of this term. However, there is enormous ambiguity as to what constitutes a high-tech sector, and which sectors generate the externalities from investment in technology which allegedly lead to suboptimal levels of investment because of problems of appropriability. For example, the agricultural sector is a sector where rates of technological innovation over the past few decades have had profound effects on levels of output, optimal size of farming units, and the ratio of capital to labour. Even in many traditional industrial sectors, such as the steel and automobile sectors, rates of technological innovation have been quite dramatic. If the government pursues some form of proactive role in promoting high technology sectors, where does it begin and where does it end? I am not aware of any compelling evidence that the externalities associated with technological innovation in some of these sectors relative to others are so great as to warrant singling out these sectors for special promotional efforts. Indeed, it may be the case that our ability, as a relatively small economy, to free-ride on investments in technological innovation that occur in other larger economies is greater with respect to the so-called knowledge-based, micro-electronic, information-intensive sectors than in other sectors, where technological innovation is important. Thus, the case for proactive intervention by government may plausibly be argued to be weaker, at least beyond supporting non firm-specific education and training and basic pre-commercial research.

David Wolfe's chapter stands, to a large extent, in interesting contrast with Professor Howse's chapter; it adopts a much more optimistic view of the role of government in promoting high technology industries, by which the author appears to mean knowledge-intensive sectors relying heavily on microelectronic technologies as inputs. Wolfe reviews various strands of literature that emphasize the importance of national systems of innovation and networks of public and private sector institutions that promote technological innovation and diffusion. He then goes on to emphasize that increasingly these networks tend to function most effectively at a regional level, in part drawing on Porter's theory of economic clusters. He concludes with a review of evolving Canadian federal and provincial (Ontario) industrial policies within this framework. As with Howse's chapter, let me note a number of reservations or questions I have about Professor Wolfe's analysis.

In emphasizing the importance of networks and network externalities, Wolfe tends to assume, without any serious qualification, that these technology-driven networks are inherently socially beneficial. However, as a growing literature in the anti-trust field reveals, there are grounds for some caution towards the uncritical endorsement of industrial networks, whatever their nature or functions. First, these networks may simply be an elaborate cover for horizontal forms of anti-competitive, welfare-reducing collusion. Second, it has been revealed in a number of industries where deregulation was thought likely to create more competitive and efficient industry structures, particularly where technological innovations exist that challenge the natural monopoly assumptions traditionally made about these industries, that these claims have often proven exaggerated. This exaggeration is a result of the scale of network economies that make it very difficult for firms to compete in only one segment of such a network. This has proven to be the case in telecommunications, airlines, ocean shipping, railways and interactive banking. In short, highly vertically integrated industry networks may raise barriers to entry for new firms, potentially reducing the vitality and rates of innovation in the industry over the longer term, and permitting anti-competitive pricing and other practices. In turn, I do not want to overstate these concerns, but simply make the point that 'networks' are not *per se* good.

Wolfe's emphasis on regions rather than nations becoming an increasingly important locus of activity, and therefore subnational levels of government playing an increasingly important role in promoting regional networks and clusters, does not address the problem raised by Howse of interjurisdictional competition for new investment. In some cases this may be socially desirable. In other cases such interjurisdictional competition may involve a zero sum form of corporate giveaways which have little or no effect on aggregate national economic growth rates (and perhaps even regional growth rates), but firms end up capturing rents from intergovernmental bidding wars. In this respect, it is worth noting that the new Internal Trade Agreement (which was signed in July of 1994 and came into effect in July 1995) imposes a rather loosely framed set of constraints on provincial governments engaging in these bidding contests for new investment. Thus, Wolfe needs to address more fully the question of when it is appropriate either from a regional or national point of view for regional level of government to engage in this kind of interjurisdictional competition.

In tracing the recent history of industrial policy in Ontario, the history in the 1970s and early 1980s seems to me to be less edifying than one might infer from the author's account. First, the Ontario Development Corporation, while no doubt supporting various new activities, was also drawn into bailing out a number of failing firms, many of which we described in our study, *The Political Economy of Business Bailouts*, for the Ontario Economic Council in 1984. These included unsuccessful attempts at bailing out firms such as Manaki Lodge and Clarke Irwin, as well as an inexplicable portfolio-type direct investment by the government in Suncor. In terms of promoting sunrise as opposed to sunset industries, the IDEA Corporation quickly became embroiled in scandals, and it appears to have achieved next to nothing in return for the large sums of money invested in so-called high-tech industries.

Wolfe refers to a recommendation of the Premier's Council which called for the creation of a Strategic Procurement Committee to promote, presumably through government procurement, the development of competitive Ontario suppliers. Again, as in some of my comments on Professor Howse's chapter, it requires noting that under the Uruguay Round Government Procurement Code, the NAFTA Government Procurement Provisions, and the Procurement Provisions in the new Internal Trade Agreement, the ability of national and subnational governments in Canada to discriminate in favour of local suppliers is now increasingly constrained as an available industrial strategy.

In reviewing the author's account of various initiatives that have been taken in Ontario over the last few years as part of its cooperative industrial policy, I find it difficult to focus on precisely what the essence of these initiatives is. Is it in the promotion of commercially usable research in particular sectors or in particular technologies that may be used across sectors? Is it the promotion of the diffusion of technological knowledge within or across sectors? Is it the redress of alleged gaps in capital markets through the provision of publicly underwritten venture capital for start-up high technology businesses? Is it the facilitation of cooperative research or production arrangements among firms which, for some reason, are unable to forge such arrangements for themselves? In other words, exactly what are the basic purposes of cooperative industrial policy against which the initiatives can be evaluated in some meaningful fashion? How would we know whether these initiatives have succeeded or failed? What constitute relevant operational and performance benchmarks, and what constitute the counterfactuals that current policies are being compared with? What systematic program or policy evaluation mechanisms does the government have in place, and how is some acceptable measure of public accountability for the expenditure of public resources on these initiatives ensured?

In suggesting that current industrial policies in Ontario entail a significant break with the past, Professor Wolfe is less than forthcoming about the extent to which the present government has persisted with the policy of past governments in bailing out failing firms, e.g. Algoma Steel, Spruce Falls Pulp and Paper Mill, Canadian Paper International, De Havilland, and Ontario Industries. In terms of aggregate spending on industrial policies, what proportion has been spent on these bail-outs compared to expenditures on cooperative industrial policies in high technology industries?

NOTE

1 See, e.g. Paul Romer, 'The Origins of Endogenous Growth Theory' (1994) 8 *Journal of Economic Perspectives* 1; Gene Grossman and Elhanin Helpman, 'Endogenous Innovation and the Theory of Growth' (1994) 8 *Journal of Economic Perspectives* 23; Pranab Bardhan, 'Economics of Development and the Development of Economics' (1993) 7 *Journal of Economic Perspectives* 129; Miranal Datta-Chaudhuri, 'Market Failure and Government Failure' (1990) 4 *Journal of Economic Perspectives* 24.

Part II

TAX POLICY AND TECHNOLOGY DEVELOPMENT

4

TAX INCENTIVES AND COMPARATIVE ADVANTAGE

Paul Halpern and Jack Mintz

INTRODUCTION

Canada relies on a tax system that not only raises revenue, but also contributes to economic and social policy. The tax system has been used to encourage specific economic activities, penalize socially undesirable behaviour and redistribute resources. Like most industrialized countries, Canada relies on several taxes to achieve its revenue-raising and economic objectives. The key components are related to the taxation of labour (the personal income tax and payroll taxes), capital (corporate and personal income taxes, capital asset taxes, sales and excise taxes on capital inputs and property taxes) and consumption (sales and excise taxes).

The Canadian tax system has been geared towards traditional industries in the primary and resource sectors which were a significant part of Canada's industrial base during the first part of this century. As a result of global economic integration in the latter part of this century, attention is being directed towards Third World countries to produce raw materials, where unskilled labour is cheapest and resources are abundant. The impact on the Canadian economy of these changes has been significant. We will present evidence below (pp. 77–81) that the relative importance of sectors in the Canadian economy is changing from the traditional resource industries to technology-based industries in manufacturing and services. However, has the set of tax incentives in the Canadian tax system changed as these economy wide changes have arisen, or has it remained unaltered reflecting an outdated social engineering posture? As we show in the next section (pp. 74–6), the tax system continues to provide incentives for industries and sectors of the economy that have not displayed growth over the recent period.

For Canada to maintain its standard of living and equality of opportunity for its citizens, a new approach is needed for taxation. The view taken here is that the most important change is to reduce incentives for traditional industries and improve the economic climate for investment, savings and risk-taking. Whether this change suggests a level playing field for all industries or a continuation of social engineering by introducing tax incentives to those industries that are currently in vogue is not addressed. Our concern is that the incentives given to traditional industries should be eliminated. However, given the inertia that is found in governments, we believe that attempts to align tax incentives with the prevailing high growth industries can lead to similar problems in the future as the high growth industries evolve to medium or even negative growth.

TAX INCENTIVES: MARGINAL EFFECTIVE TAX RATES

Historically, Canada, as a resource rich country, has directed its tax system to support agriculture, fishing, forestry, mining, and oil and gas industries. Despite tax reforms, the current tax system offers a host of tax incentives and special exemptions for primary and resource firms, including special write-offs under the income and mining tax laws, federal and provincial sales and excise taxes, property tax and customs tariffs. In Appendix I (pp. 85–9) we present a listing of exemptions to GST and Customs and Duties for the fishing and farming sector. A comprehensive list of special exemptions directed towards mining, oil and gas and forestry industries is found in Chen, Mintz, Scharf and Traviza (1995).

Another way to look at the tax incentives provided through the tax law is to measure effective tax rates. Effective tax rates assessed on corporations are very important in terms of decisions made by corporations. The tax rates will impact the amount of long-term investment decisions, and the relative use of capital and labour within the organization. In this section we identify the effective tax rates assessed on a set of eleven industries. We calculate the marginal effective tax rates on capital, labour and overall cost of production. The marginal effective tax rate (METR) is defined as the amount of taxes paid on income earned by the incremental unit; in the case of capital the unit is the incremental investment, whereas in the case of labour it is the incremental worker. For the overall cost of production we use only capital and labour inputs and the METR reflects the additional taxes paid as a percentage of the cost of incremental production.

In Table 4.1, the results of the estimation of the METR for eleven industries are presented. (A technical description of the derivation of the METRs for capital, labour and overall cost of production is presented in Appendix II, pp. 89–91.) Beginning with the capital component, we have identified the METRs for large and small firms, along with the weighted average of the two components referred to as combined. The last column presents a simple average of the industry marginal effective tax rates.

Table 4.1 Marginal effective tax rates (METRs, %)*

	METR on capital				
	Large firms	*Small firms*	*Combined*	*METR on labour*	*Overall METR on cost of production*
Agriculture, Fishing and Forestry	39.77	13.47	19.01	16.48	18.19
Mining†	−6.89	−3.11	−6.05	15.81	2.60
Oil and Gas†	63.08	53.21	60.88	19.91	47.51
Manufacturing	27.92	11.55	23.74	16.52	20.51
Construction	55.59	14.84	23.71	16.61	19.93
Trade and Storage	23.33	5.11	17.31	16.58	16.95
Communication	14.97	1.63	14.42	16.66	15.53
Utility	24.42	5.26	22.67	16.23	20.94
Trade	53.52	16.59	36.16	16.62	21.25
Service	20.21	3.53	11.87	16.59	15.00
Financial	35.81	4.48	33.36	16.69	24.64
Average	31.98	11.51	23.37	16.79	20.28

Notes: * The METR is calculated with industry-specific debt to asset ratios.
† The estimates of METR for Mining and the Oil and Gas industries treat provincial mining tax and royalties as tax. By excluding these charges from our estimates, the METR on capital for Mining and Oil and Gas will be −8.11 per cent and 8.66 per cent, respectively; the METR on labour will be 18.80 per cent for Oil and Gas and, the METR on cost of production will be 1.29 per cent and 11.66 per cent for Mining and Oil and Gas, respectively.

An important part of our calculations is the inclusion of royalties paid to governments for the use of public lands and deposits. In the calculations we have treated the provincial mining tax and royalties as tax. This would be appropriate if it is viewed that governments could collect resources royalties with a rent tax that would not affect the incremental cost of production or the cost of capital.[1] Any other type of royalty besides a rent tax would be distortionary and thus be viewed as a tax. On the other hand, most governments assess mining profit taxes (profits being defined as the return on capital gross of interest expenses) or royalties on the value of production. The latter form of royalty is often used by the private sector for assessing payments made by resource companies for the use of property owned independently. If we were to consider provincial oil and gas royalties and mining profits taxes as a business expense, rather than a tax, the METRs are affected for mining and oil and gas.

In all sectors, the METR for small firms is much smaller than for large firms, primarily as a result of the small business deduction at the federal level and favourable provincial tax rates for small firms. The METR for small firms relative to large firms is smallest for the mining and oil and gas industry, and much larger for all other industries. For example, in the communications sector the METR for small firms is 11 per cent of the METR of the large firm. Looking at the combined values, the spread in METRs is very large, ranging from a subsidy value of −6.05 per cent for mining to a high of 60.88 per cent for oil and gas. The next highest is the financial sector. However, if royalties etc. are treated as expenses, the METR for mining is −8.11 per cent and for oil and gas, 8.66 per cent.

If royalties and provincial taxes are considered as expenses, the simple average of the METRs is approximately 19 per cent. There are five sectors which have METRs below this average: mining, oil and gas, trade and storage, communications, and services. The first two sectors have the greatest deviation. If the royalties and provincial taxes are included as tax, the average METR is approximately 23 per cent; and the same sectors, except for oil and gas, are below the average.

The METR for labour is presented in the middle of Table 4.1. This is composed of the charge for Canada Pension Plan, Unemployment Insurance premiums, provincial payroll taxes and personal taxes on labour income. It is assumed that only part of the taxes are shifted forward as higher costs on firms (30 per cent as suggested by Dahlby (1992)). The resulting METRs are not dramatically different across the sectors. The only outlying observation is for the oil and gas sector, which faces a higher METR on labour since royalties which apply to revenues affect not only capital costs but also wage costs. Adjusting for provincial royalties by treating them as a tax, the METR on labour falls to 18.80 per cent, still the highest value.

The overall METR on cost of production is the last entry in Table 4.1. Including royalties as tax, the average METR is approximately 20 per cent. The METRs range from a low of 2.60 per cent for mining to a high of 47.51 per cent for oil and gas. Adjusting for the royalties, the METR for oil and gas becomes 11.66 per cent and the overall average is approximately 17 per cent. Regardless of how royalties and provincial taxes are handled, it is obvious from this analysis that mining has an extremely low METR. Other sectors which appear to be subsidized through a low METR are services, communications, trade and storage.

One surprising result is that the agriculture, forestry and fishing sector does not appear to have a low METR. However, these calculations do not include other tax benefits and subsidies to this sector, many of which have been presented in Appendix I (see pp. 85–9).

Therefore, based on the overall METRs on the cost of production, the tax system has provided an incentive to certain sectors in the economy. The question we address is whether these sectors are also those in which there is significant growth and whether they are important in the Canadian economy.

COMPARATIVE ADVANTAGE OF INDUSTRY SECTORS IN THE CANADIAN ECONOMY

There are a large number of sectors in the Canadian economy which are important in both domestic and international production and trade. The explanation of the evolution of the importance of these sectors and their relative growth over time is a complex issue which depends upon a number of economic factors, including government subsidies which influence the size of the marginal effective tax rate (METR) applied to each sector. It is not the purpose of this chapter to try to relate the observed growth and importance of certain sectors in the economy to a set of economic variables, including the METR. In fact, we take the observed growth of these sectors and relate this to government policy as measured by the METR.

The choice of sectors used is constrained somewhat by our measurement of the METR which was described above (pp. 74–6). Since this calculation is key to the analysis, we try to choose sectors for which relative growth rates are measured to overlap with the sectors used in the METR analysis. Unfortunately, additional problems arise since the sectors used by different data collection agencies are not consistent. We have attempted to match as best as possible the definitions of the sectors used in the METR material as well as in the measurement of sector growth.

In order to assess the growth of sectors we use four different measures: non-export related measures, such as growth in employment and growth in GDP; and export related measures, such as growth in exports and the change in Canada's Revealed Comparative Advantage (RCA). In Table 4.2 we consider the annual growth in employment by sector for both goods and services over the period 1970 to 1993. The employment growth in the goods sectors is very small; and in the case of the combined group of agriculture, mining, and oil and gas, the growth rate is negative. From Table 4.2 it is clear that employment growth is much higher in the services sectors than in the goods segment. The largest growth rate is in community business and personal services at 3.55 per cent, and the lowest in the combined group of transportation, communications and utilities at 1.16 per cent.

The annual growth rate in GDP over the period 1970 to 1993 and the individual sector's share of total GDP over the period for 1970 and 1993 are presented in Table 4.3. Looking first at the goods segment, the manufacturing sector has the highest share in 1993 at 17.8 per cent (down from its 1970 value of 20.6 per cent). The other sectors, while not having as large a share of total GDP as construction, still had a reduction of their share from 1970. The annual growth rates over the period for these sectors ranged from 1.0 per cent to 2.4 per cent.

Table 4.2 Average annual change in employment by sector, 1970–93 (thousands of persons)

	1970	*1993*	*Average annual change (%)*
Goods:			
Agriculture, Fishing and Farming, Oil and Gas, Mining*	729	699	−0.18
Manufacturing	1,768	1,800	0.08
Construction	467	660	1.52
Services:			
Transportation, Communication, and Utilities	698	910	1.16
Trade	1,328	2,138	2.09
Finance, Insurance and Real Estate	379	768	3.12
Community, Business and Personal Services†	2,040	4,552	3.55

Source: *Historical Labour Force Statistics* (71–201) (Ottawa: Statistics Canada).
Notes: * Statistics Canada provides employment data for the two categories of Agriculture and Other Primary; however, because our METR calculations combine Agriculture with Forestry and Fishing, but separate Other Primary into Mining and Oil and Gas, we are forced to aggregate all the employment figures into one category.
† No corresponding METR is available.

The most interesting observation in Table 4.3 is for the services segment. As a share of GDP, finance, etc. and community business and personal services had the greatest shares of GDP in 1993 (and these shares increased from their 1970 values). All of the sectors in the services segment, apart from 'transportation and storage' and 'government services', had an increase in their share of GDP. The largest annual growth is found in the communications sector at 7.1 per cent, which with a 3.8 per cent share of GDP in 1993 is an important sector in the Canadian economy. All of the other sectors in which the shares in GDP increased over the period 1970 to 1993 had annual growth rates above 3.0 per cent. In fact, the government services growth rate had the lowest observed growth rate in the services segment and was closer to the values found in the goods segment.

Turning to the export side, in Table 4.4 we consider the average annual increase in exports for both goods and services over the period 1977 to 1993. The starting date reflects the availability of data presented in the source. In the goods group, consumer goods, machines and equipment and autoparts have the highest annual growth rate at 8 per cent; with an annual growth rate of 4 per cent in the other sectors, agriculture *et al.*, industrial goods and energy products. In the services sector, business services increased annually by 8 per cent, and transportation and storage

Table 4.3 Average annual increase in GDP and change in share of total GDP at factor cost by industry (in millions of 1986 dollars)

	GDP 1970	*GDP 1993*	*Average annual increase (%)*	*1970 share total GDP (%)*	*1993 share total GDP (%)*	*Change GDP share (%)*
Goods:						
Agriculture, Fishing, Forestry, Trapping and Related	10,766	16,193	1.8	4.2	3.1	−1.02
Mining	17,227	21,635	1.0	6.6	4.2	−2.46
Manufacturing	53,319	92,270	2.4	20.6	17.8	−2.73
Construction	16,684	27,626	2.2	6.4	5.3	−1.10
Services:						
Transportation and Storage	13,825	2,453	12.5	5.3	4.7	−0.59
Communication	4,045	19,577	7.1	1.6	3.8	2.22
Finance, Insurance and Real Estate	3,327	38,726	24.3	12.8	16.9	4.03
Community, Business and Personal Services*	54,766	115,453	3.3	21.1	22.3	1.19
Trade	28,404	62,738	3.5	11.0	12.1	1.17
Utilities	6,686	16,335	4.0	2.6	3.2	0.58
Government Services*	20,239	33,785	2.3	7.8	6.5	−1.28
Totals	259,234	517,405	3.1	100.0	100.0	

Source: *Canadian Economic Observer* (Ottawa: Statistics Canada, 1993/94).
Note: * No corresponding METR is available.

by 3 per cent. Although this is a fairly short time period, one interesting result is the growth of business services exports.

One final indicator of the importance in the international trade area of various sectors in the Canadian economy is Revealed Comparative Advantage (RCA). The RCA for any good or set of goods is composed of two ratios. The numerator is the proportion of Canadian exports to total exports for a particular good. The denominator is the proportion of world trade made up by the same good or set of goods. For example, if we are interested in the mining sector, and the mining sector is 10 per cent of world trade and 20 per cent of Canada's trade, then the RCA will be greater than unity. If the RCA is greater than unity, the good is more important in Canada than in the rest of the world.

Unfortunately, the RCA does not provide complete coverage of all the important sectors involved in international trade. Since there is as of yet little standardization of definitions in the services sector, RCA does include

Table 4.4 Average annual change in exports by sector, 1977–93 (in millions of 1986 dollars)

	1977	*1993*	*Average annual increase (%)*
Goods:			
Agriculture, Fishing and Forestry Products	20,398	37,096	4
Manufacturing*	24,451	84,665	8
Mining†	15,658	29,562	4
Oil and Gas‡	8,650	17,060	4
Services:			
Financial, Communications and Business Services§	2,680	9,666	8
Transportation and Storage	3,687	5,584	3

Sources: Canadian Economic Observer (Ottawa: Statistics Canada, 1993/94, 11–210); *Canada's International Transactions in Services* (Ottawa: Statistics Canada, 1992/93, 67–203).

Notes: * The components of manufacturing exports provided here include figures for consumer goods, machines and equipment, and autoparts.

† The components of mining include Statistics Canada data for exports of Industrial Products which is their classification for minerals and its semi-processed by-products.

‡ These figures for oil and gas include coal, whereas coal is not included in the oil and gas calculations for the METR.

§ The data from Statistics Canada aggregates all services exports into the broad category of Business Services; amongst other things, this includes exports of financial and communications services, as well as the components of our calculation for the Services METR.

the services sector. For Canada, as noted in Table 4.4, services exports have grown very rapidly, and not including them in the RCA calculations will increase the RCA values for the traded goods sectors if the services sector is more important in Canada than in the rest of the world.

However, even with this caveat, the RCA analysis does provide us with an indication of the importance in the Canadian economy of standard traded goods. While RCA values can be identified at a very disaggregated level, we have been forced to aggregate, to reflect as closely as possible the definitions of sectors used in the METR calculations.

In Table 4.5, the striking observation is the decrease in the RCA for the food/agriculture sector falling to 1.222 in 1992 from a value of 9.166 in 1970. Also, the RCA for ore fell from a value of 2.880 in 1970 to 0.227 in 1992. The RCA values for both of these sectors demonstrate their reduced importance in Canada's trade. The only sector to increase in importance is the energy sector, where it has a current value just in excess of unity.

Table 4.5 Changes in Canada's Revealed Comparative Advantage in goods by sector (1970–92)

	RCA 1970	*RCA 1992*
Agriculture and Fishing*	9.166	1.222
Manufacturing	1.020	0.901
Mining†	2.880	0.227
Oil and Gas‡	0.680	1.097

Source: International Trade Data Collection (Chelem CD Rom).
Notes: * Unlike our Agriculture, Fishing and Forestry METR, this data excludes forestry products, but includes processed food products.
† This category represents export figures for ore products, which are mostly composed of minerals.
‡ This includes coal exports which are excluded from the METR for Oil and Gas.

Two general conclusions arise from the data presented in Tables 4.2 to 4.5. First, growth in employment and GDP suggests that services are growing faster than the goods sector; services are becoming more important in the Canadian economy as measured by their share of GDP. The importance of services is also observed in growth in exports by sector. Second, looking at goods alone, traditional industries such as agriculture, mining, and even manufacturing have shown lower growth rates and a declining share of GDP. Since the last category is very inclusive, the observations conceal the growth of certain sub-sectors, such as autoparts and machinery and equipment.

CONSISTENCY OF TAX CONCESSIONS AND SECTOR GROWTH

Table 4.6 presents a summary of the marginal effective tax rates (METRs) and the measures of growth for the sectors identified in the METR analysis. A striking inconsistency becomes apparent when all the data is aggregated in this form: the financial services sector, which boasts the highest results in all applicable measures of growth, is subjected to the highest METR; while mining, with its almost negligible METR of 1.29 per cent, has produced very poor results in those same measurements (including a drop in its RCA). Likewise, agriculture, fishing, and forestry, with its relatively low METR and extensive legislative tax exemptions (see Appendix I), also displayed negative employment growth,[2] small increases in GDP, a fall in its share of GDP, relatively small figures for export growth, and a considerable drop in the RCA index from 9.166 to 1.222.

Outside the parameters of the above two extremes, the results turn slightly 'grey'; sectors like communications or services, both exhibiting

Table 4.6 Summary of METR and various measures of sector importance*

Sectors defined by METR analysis	*METR overall*	*Employment % growth*	*GDP % growth*	*Share in GDP +/−*	*Export % growth*	*Change in RCA*
Agriculture, Fishing and Forestry	11.79	−0.18	1.8	−	4	−
Mining	1.29	−0.18	1.0	−	4	−
Oil and Gas	11.66	−0.18	n.a.	n.a.	4	+
Manufacturing	20.51	0.08	2.4	−	8	0
Construction	19.93	1.52	2.2	−	n.a.	n.a.
Transportation and Storage	16.95	1.16	2.5	−	3	n.a.
Communications	15.53	1.16	7.1	+	8	n.a.
Utilities	20.94	1.16	4.0	+	n.a.	n.a.
Trade	21.25	2.09	3.5	+	n.a.	n.a.
Services	15.00	3.12	n.a.	n.a.	8	n.a.
Financial	24.64	3.12	4.3	+	8	n.a.
Source:	Table 4.1	Table 4.2	Table 4.3	Table 4.3	Table 4.4	Table 4.5

Note: * If two or more entries have the same value, then the sectors have been aggregated or are not available with the same definition used in the METR analysis.

good results in the growth measurements, are already subject to reasonably moderate METRs. Hence, aside from inducing a debate over the direction of or need for government tax strategy, a call for more reasonable tax treatments of our growing service industry may, at least in the case of these two sectors, be unnecessary under our current tax regime. None the less, the strong relative performance of the components of the communications and services[3] sectors must not be overlooked. Other 'grey' areas include manufacturing, where comparatively low employment and GDP growth rates, and a drop in share of GDP, are mixed with a neutral change in RCA, high export increases, and a high METR. Ambiguous results like these are also found in the oil and gas sector, with its low METR, low aggregated employment figures, and positive change in RCA.

Notwithstanding these unclear results, there are still two conclusions which can be affirmed from the above data. The first is that low METRs do not appear to work as an effective buttress against fundamental changes in the world's makeup of comparative advantage. Clearly in Canada's case, the low METRs and the legislated tax breaks to agriculture, mining, forestry, and fishing, have not offset the low wage competition from resource-rich developing countries or any of the other factors accounting for the relatively poor performance in the resource sector over the last fifteen to twenty years. On the contrary, they merely reflect continued government support for failing industries and lost revenue. The second conclusion to be drawn from the data points to the fact that the financial sector has been able to perform so remarkably well despite the high tax burdens. Because there is little uncertainty that marginal tax rates do affect marginal investment decisions, and in turn growth and profits, the lofty METR of 24.64 per cent on financial services could account for a considerable loss of potential growth in this industry beyond its already impressive figures.

Therefore, although we recognize that these conclusions are largely impressionistic, we see no rationale for the inconsistencies in the METR/growth rates analysis. That is to say, we see no logical economic reason to subject the healthy sectors of the economy to the highest METRs and the least competitive sectors of the economy to the lowest METRs. In the next section we summarize what all this implies for Canada's current and future tax regime.

CONCLUSIONS: TAX POLICY MUSINGS

Canada's tax system is ill-equipped to accommodate new economic trends as we move closer to the twenty-first century. International economic integration requires governments to choose policies that would maintain economic growth by improved investment and export performance. Policies should be directed towards technology, education and sustainable development to enhance Canada's distinct technological advantages in providing goods and services to the world market. Our substantial international indebtedness, now over 40 per cent of GDP, must be curtailed if we are to improve our standard of living.

What Canada needs now is a tax system that provides a healthy environment for investment, savings, risk-taking and the acquisition of technology and skills. This would require a shift in priorities that favours consumption and natural resource production towards one that is less punitive towards investment, savings and risk-taking.

There are two approaches that could be used to improve Canada's tax structure. The first is to maintain the current tax structure, but reduce its discrimination in favour of some activities and against others. In other words, one can 'level the playing field' by evening out effective tax rates on capital and the cost of production. This has been the theme of tax reform of the later 1980s. To accomplish a level playing field going beyond the recent tax reform measures, some of the policies that would need to be adopted include the following:

1 Adopt depreciation rates close to economic depreciation rates so all types of capital will be taxed at a similar rate.

2 Tax all forms of imputed income or match deductions for expenditures on exploration and development, research and development, etc. Put agriculture and fishing on an accrual rather than cash basis for measuring income.
3 Eliminate fast write-offs such as investment tax credits in the Atlantic provinces, flow-through shares, and accelerated depreciation for manufacturing and mining.
4 Provide full loss offsetting by refunding losses or allowing firms to carry forward losses at a full rate of interest. This would not penalize risk-taking or growing firms.
5 Eliminate sales and excise tax on capital inputs since they apply more heavily on capital with shorter lives. Eliminate sales taxes on other business inputs so that goods and services are taxed at similar rates.
6 Impose capital taxes on all assets measured according to economic rates of depreciation.

These examples of tax policies needed to achieve a neutral tax system have not been, and are never likely to be, fully adopted. In part, it is impossible to measure all form of imputed income and economic depreciation properly. Full loss offsetting would be difficult if it leads to significant revenue losses for the government, in part due to evasion.

Recently, there has been debate to completely change the tax system. In the United States, for example, flat rate tax plans have been resurrected by the Republicans. The simplest of these plans calls for a flat rate tax applied to the consumption (cash flow) of individuals and companies (e.g. income net of capital expenditures or savings). Similar proposals have been suggested in Britain (the Meade Report) and Canada (the MacDonald Report).

The consumption tax system has attracted attention as a result of taxpayer frustration with a complex tax structure in today's world, especially the income tax. However, the consumption system is an entirely different type of tax. Although it avoids the tricky problems of defining depreciation and imputed income, the consumption tax implies that normal return on investments is exempt from taxation (only consumption arising from above normal returns is taxed). It also implies that the effective tax rate on incremental investment projects is exempt from taxation altogether. One could argue this is a healthy result. As pointed out by the Meade Report (Institute for Fiscal Studies, 1978), for example, taxing the return on savings and investments results in excess taxation of future consumption or production relative to current consumption or production. The reason for this is that personal earnings are already taxed once. If the earnings are immediately consumed, no further income tax is owing. If, on the other hand, earnings are saved for future consumption, the return on savings is subject to a tax. Thus, savers are more highly taxed than consumers.

The idea of not taxing capital investment, at least at the margin, has not received wide popular support to date. However, it has influenced the development of world-wide VAT (value-added tax) systems which are a form of consumption taxation. (VAT is applied to a business's revenues net of expenditures on capital and other goods without a deduction for wage expenditures.) In recent years, a number of countries have converted their VAT from an 'income' base to a 'consumption' base by expensing net capital purchases rather than deducting depreciation for capital costs (e.g. Argentina and Egypt). Some countries, such as New Zealand, have reduced income taxes in favour of a broad-based VAT. Perhaps, therefore, the consumption tax system is being introduced slowly, in the form of the VAT, as a way of eliminating the current bias of the tax system against investment.

APPENDIX I: LIST OF GOODS RELATED TO AGRICULTURE AND FISHING ENTERING CANADA DUTY FREE

- Ploughs (all types)*
- Disc harrows (all types)*
- Other types of harrows, cultivators, weeders and hoes*
- Seeders, planters and transplanters (not operated manually)*
- Manure spreaders/fertilizers (not operated manually)*
- Agricultural type domestic rotary tillers*
- Cutter bars (mowers) specifically for tractors used in agriculture (same items not used in agriculture are taxed between 2.5 per cent and 9.2 per cent)†‡
- All haymaking machinery
- Straw or fodder balers, including pick up badders
- Combine harvester-threshers
- All other threshing machinery
- Root or tuber harvesting machines
- Forage harvesters
- Agricultural (or horticultural) machines for cleaning, sorting, or grading eggs, produce, fruit or other agricultural produce (same items not used in agriculture are taxed between 2.5 per cent and 9.2 per cent)†‡
- Milking machines*
- Cheese-making machines*
- Machinery for preparing animal feeding stuffs*
- Poultry incubators and brooders*
- Crop protection and preparation machinery*
- Barn and barnyard machinery*

- Feeders or watering troughs*
- Fanning mills for agriculture*
- Fish skinning, deheading, or grinding machines
- All new tractors* and cabs used in agriculture
- Spraying lorries (spraying vehicles) used in agriculture for spreading dry or liquid fertilizer or sludge (same equipment not used in agriculture is taxed between 2.5 per cent and 9.2 per cent)†
- Fishing vessels, factory ships and other vessels for the processing or preserving of fishery products

Notes: * All necessary parts also enter duty free.
† Rate varies depending on country of origin.
‡ These items do, however, enter duty free from the US due to the FTA.

List of Statutory Concessionary Provisions for agricultural and fisheries imports

Although the following list of goods are normally subjected to any combination of MFN tariffs, General Preferential tariffs, and/or US tariffs, Schedule II of the Statutory Concessionary Provisions explicitly exempts these goods from customs if they are imported specifically for use in agriculture or commercial fishing; the number codes in parentheses pertain to the tariff item codes, and can be referenced for the detailed specifications of the goods.

Agriculture

- Binder or baler twine used for farm produce
- Electric generating sets and rotary converters for use on the farm for farm purposes only (85.02)
- Fuel tanks (7310.10.10; 73.10.29.10)
- Spark-ignition reciprocating or rotary internal combustion engines and parts (84.07; 84.09)
- Generators (85.01) and parts (85.03)
- Electric accumulators (85.07)
- Switchboards and panels (85.37) and parts
- Windmachines/fans (8414.59) and parts (8414.90.50)
- Articles for use as markers in crop spraying or crop dusting operations, designed to be ejected from aircraft
- Phosphates of calcium for use in the manufacture of animal or poultry feeds (2835.26.00)
- Filtering elements (5911.90.90) for use with milking machines

Fishing

- Net floats, spat collectors, and collector holders (3926.90.90)
- Marker buoys and beacons of non-wood products (8907.90.10)
- Fishing nets and netting (56.08)
- Knitted netting used for fishing (60.02)
- Devices for keeping nets open and swivels (7326.90.99)
- Articles and materials for use in the manufacture or repair of lobster traps or scallop drag nets
- Spark-ignition reciprocating or rotary internal combustion piston engines (as listed for agriculture)
- Compression ignition internal combustion piston engines (diesel or semi-diesel engines) (8408.10.90) and parts (8409.99.93)
- Articles and materials used in the manufacture and repair of engines or parts thereof for boats used in commercial fishing operations

List of GST provisions for fishing and farming

Fishing

- GST does not apply to royalties paid on or 'rights to explore/exploit' any water or fishing resources
- Bait used in commercial fishing is a GST zero-rated fishing product
- GST zero-rated fishing equipment purchases include the following:

  1 Fishing boats
  2 Nets
  3 Automatic baiters
  4 Mechanical net washers
  5 Netpens
  6 Fish feed

- The Input Tax Credit exempts fisherman from the GST on costs incurred for the following goods/services:

  1 Transportation of final products to purchaser
  2 Petroleum products
  3 Fishing vessel repairs, engine and electrical repairs, maintenance materials
  4 Fuel, small tools, utilities
  5 Building rentals related to fishing operations
  6 Construction materials (such as lumber, nails, paint, etc. used in structural improvements, additions or general repairs to buildings related to fishing business)

7 Fishing gear like knives, gloves, and rubber or oilskin clothing purchased for commercial fishing

Farming

- Any rent payments (or portion thereof) on farmland paid in the form of products from the land are not subjected to GST
- Purchase of farmland is exempt from GST in the form of an input tax credit; only the proportion of land used for commercial farming purchases qualifies for the exemption
- Other items/services which qualify for GST exemption through the Input Tax Credit to agriculture (i.e. GST refund) include the following:

1 Personal computers
2 Veterinary fees and drugs
3 Ceiling-mounted balm fans
4 Building rental
5 Fuel, lubricating oil and grease, utilities
6 Machinery repairs and maintenance
7 Building materials (such as lumber, nails, siding, paint, etc.) used in the construction, expansion, or repair of farm buildings
8 Flower seeds (if the final product, flowers, are resold)
9 Snowploughing equipment (if snowploughing services are provided)
10 Transportation costs of moving animals and produce from the farm to the market/consumer
11 Percentage of electricity and heating fuel used for maintenance of livestock

- Because the agricultural sector, by and large, produces goods ultimately destined for GST exempt human food consumption, almost all sales of final farm products are zero-rated; however, special GST provisions do exist for non-food agricultural items like unprocessed wool or tobacco
- Farming equipment which is zero-rated (GST exempt) at the point of purchase includes the following:

1 Bulk fertilizers
2 Large farm tractors
3 Combines, ploughs and headers
4 Pick-up trucks
5 Field, fruit and vegetable harvesters
6 Farm bulk milk coolers and tanks
7 Rock pickers
8 Discers and rock weeders
9 Seeders and planters

10 Swathers and windrowers
11 Mower conditioners, hay balers, hay rakes, hay tedders, hay conditioners/crushers/crimpers, and swath/windrow turners
12 Combination discer-cultivators
13 Complete feeds and complete feed supplements sold in bulk
14 Sales of quotas between farmers for zero-rated products

Sources

Dancey, Kevin, Kesler, Rick H., Puthon, Keri, Resendes, Ray, *A Guide to the Goods and Services Tax* (Ottawa: CCH Canadian Limited, 1991).
Government of Canada, *Goods and Services Tax: Information for Farmers* (Ottawa: Revenue Canada, 1990).
Government of Canada, *Goods and Services Tax: Information for Fishermen* (Ottawa: Revenue Canada, 1990).

APPENDIX II: METHODOLOGY FOR ESTIMATING MARGINAL EFFECTIVE TAX RATE

The marginal effective tax rates presented in this chapter are related only to the case of investments made by firms that pay full taxes. Estimates of marginal effective tax rate (METR) on capital, labour and overall cost of production were made for eleven industries.

METR on capital

1 User cost of capital

The theoretical assumption for our estimates is that the firm maximizes its value of cash flows by investing in capital until the marginal revenue of product (MRP) is equal to the cost of capital, or the user cost of capital. Mathematically, the user cost of capital for depreciable assets is defined as:

$$\mathrm{MRP} = (r + d)[1 - uZ + t(1 - u)/(a + r + p)]/(1 - u),$$

where r = real interest rate, d = economic depreciation rate, u = combined federal and provincial corporate income tax rate, Z = present value of tax depreciation, t = provincial capital tax, a = tax depreciation rate, and p = inflation rate.

As for inventories with the FIFO accounting method, the formula is:

$$\mathrm{MRP} = (r + up)/(1 - u) + t.$$

In the case of land, we have:

$$\mathrm{MRP} = r[1 + t(1 - u)/(r + p)]/(1 - u).$$

2 *METR*

The METR on a given type of capital is defined as the proportional difference between the before-tax rate of return required by a firm (Rg) and the after-tax rate of return required by an investor (Rn). Rg is the difference between the MRP (or user cost, in equilibrium) and economic depreciation. The after-tax rate of return is the weighted average of the return to debt and equity securities held by the investor. Thus, the effective tax rate (t) is defined as:

$$t = (\mathrm{Rg} - \mathrm{Rn})/\mathrm{Rn} \text{ and } \mathrm{Rn} = bi + (1 - b)r - p,$$

where b = debt to asset ratio, i = interest rate, and r = rate of return on equity. Canada is assumed to be an open economy, whereby the after-tax rate of return is determined by the world capital market.

3 *Aggregation*

The effective tax rate for a given industry is the proportional difference between the weighted average of before-tax rate of return by asset type and the after-tax rate of return which is the same across asset type within the industry. That is, the METR for industry t_i, is calculated as follows:

$$t_i = (S_j \mathrm{Rg}_{ij} w_{ij} - \mathrm{Rn}_i)/\mathrm{Rn}_i,$$

where j denotes asset type (i.e. investments in buildings, machinery, inventories, land, and exploration and developments), w_{ij} denotes the weight of asset type j in industry i.

METR on labour

To calculate METR on labour, we assume that for a particular sector the elasticity of labour supply is infinite, and the gross-of-tax demand for labour decreases as the cost of labour increases. Since payroll taxes differ by type of worker (income level), the tax bill depends upon the composition of the workforce. For this analysis we have assumed this composition to be constant for a given industry, and the average earnings in the industry was used for estimating effective payroll tax rate and personal income tax rate. Assuming that a third of payroll and personal income taxes were borne by employers, the METR on labour was estimated as a third of the sum of effective payroll tax rate and personal income tax rate.

METR on cost of production

The cost of production in our calculation consists of cost of capital and cost of labour only. That is, material and other inputs were ignored. By applying the augmented Cobb–Douglas production function, the METR on cost of production (T) is defined as:

$$T = (1 + t_k)^{ak}(1 + t_l)^{al} - 1,$$

with t_k and t_l as METRs on capital and labour, respectively; and, ^{ak} and ^{al} as shares of capital and labour in the cost of capital.

ACKNOWLEDGEMENTS

We wish to thank Ben Cherniavsky and Duanjie Chen for very capable research assistance. This is the original paper presented at the 1995 Conference.

NOTES

1 An example of a rent tax is a cash flow tax (see Institute for Fiscal Studies, 1978 – the Meade Report). A cash flow tax applies to revenues net of current and capital expenditures (no deduction is permitted for interest expense). British Columbia uses a cash flow tax for assessing mining royalties.

2 According to Statistics Canada's categories for employment growth, it is in fact only the agriculture sector which showed negative growth in employment (−12.5 per cent) during the 1970 to 1993 period. Employment in the other primary industry categories, comprising mining, oil and gas, forestry and fishing, showed a small increase of 15.7 per cent, well below the overall employment growth of 36 per cent during this period. This discrepancy is not large enough to affect the conclusions we can draw from this data.

3 Includes employment, advertising, accounting, computing, architecture, law, managerial consulting and other business service firms.

BIBLIOGRAPHY

Bellassa, Bela, 'Trade Liberalization and Revealed Comparative Advantage', *The Manchester School of Economics and Social Studies*, Volume 33, May 1965.

Canada, *Report of the Royal Commission on the Economic Union and Development Prospects for Canada* (the MacDonald Report) (Ottawa: Supply and Services, 1985).

Chen, D., Mintz, J., Scharf, K. and Traviza, S., *Taxation of Virgin and Recycled Materials: Analysis and Policy* (Winnipeg: Canadian Council of Ministers of the Environment, 1995).

Dahlby, Bev, 'Taxation and Social Insurance', in Richard Bird and Jack Mintz, eds, *Taxation to 2000 and Beyond* (Canadian Tax Paper No. 93: 110–56) (Toronto: Canadian Tax Foundation, 1992).

Eden, Lorraine and Molot, Maureen, 'Comparative and Competitive Advantage in the North American Trade Bloc', *Canadian Business Economics*.

Halpern, Paul and Mintz, Jack, 'Taxation and Impact on Market Structure', in Richard Bird and Jack Mintz, eds, *Taxation to 2000 and Beyond* (Canadian Tax Paper No. 93: 214–68) (Toronto: Canadian Tax Foundation, 1992).
Institute for Fiscal Studies, *The Structure and Reform of Direct Taxation*, Report of a Committee Chaired by Professor J. E. Meade (London: Allen and Unwin, 1978).

5

EVALUATING R&D INCOME TAX INCENTIVES

Some lessons from the Australian experience

Gordon J. Lenjosek

INTRODUCTION

In 1993, the Australian Bureau of Industry Economics published the results of an evaluation it undertook of income tax assistance for research and development (R&D) in Australia.[1] The evaluation findings were based in large part on data obtained from a survey of R&D performers. That evaluation considered, among other things, the impact of the Australian R&D tax concession on corporate R&D spending in Australia and on the innovativeness and competitiveness of Australian firms. The evaluation also assessed the cost-effectiveness of the R&D tax concession in promoting incremental R&D (i.e. additional spending on R&D that is attributable directly to the tax concession) and the spillover benefits associated with incremental R&D.

More recently, the Canadian government announced that it is undertaking work of a similar nature. The Department of Finance and Revenue Canada are conducting a joint evaluation of the performance of the federal income tax incentives for scientific research and experimental development (SR&ED) in relation to the objectives established for them by the federal government and in keeping with Treasury Board guidelines for evaluations.[2] The Department of Finance is the federal department primarily responsible for providing the Government of Canada with analysis and advice on matters of tax policy relating to SR&ED. Revenue Canada is the federal department responsible for administering the SR&ED income tax provisions. The performance of the SR&ED tax incentives is being assessed in terms of their relevance, impacts and effects, and cost-effectiveness.

The nature and conduct of the Australian evaluation may provide some valuable insights for the Canadian evaluation study. For this reason, this chapter outlines the Australian R&D tax concession and discusses the Australian evaluation work, i.e. its objectives, the methodologies employed, the results obtained and some problems encountered. Some lessons that might be drawn from the Australian experience are provided based on this review. There are also two appendices to this chapter (see pp. 102–10). The first reviews key elements of the federal and provincial income tax incentives for R&D in Canada. The second provides an international comparison of the relative attractiveness of the R&D tax incentives provided by the G-7 countries and Australia.

GOVERNMENT SUPPORT FOR R&D

The OECD defines R&D to comprise creative work undertaken on a systematic basis in order to increase the stock of knowledge, including knowledge of man, culture and society, and the use of this stock of knowledge to devise new applications. This work may take the form of basic research, applied research or experimental development.[3]

R&D produces technology, a form of knowledge that is used to enhance the productivity of factors of production. Economic theory indicates that technological progress, primarily through its impact on labour productivity, is a key determinant of the longer-term growth of an economy. The key economic rationale for governments to assist R&D is that the benefits of R&D spill over or extend beyond performers themselves to other firms and sectors of the economy and the value of these benefits is not fully captured by the R&D performer. These 'spillover benefits' mean that, in the absence of government support, firms would likely perform less R&D than is desirable from the economy's point of view, i.e. markets fail to allocate an efficient or socially optimal quantity of resources to the performance of R&D. Furthermore, in an international and strategic context, support policies for R&D enhance international competitiveness by attracting R&D capital to a country and promoting knowledge-based jobs and economic growth.

In response to this market failure and for reasons of international competitiveness, most countries typically provide both tax and non-tax incentives for R&D. Specific incentives within these general categories can take many different forms. For example, income tax support may be provided through various types of accelerated or bonus deductions, tax credits or incremental versions of those deductions or credits. Non-tax support may be in the form of patent protection, R&D conducted by government laboratories and related establishments, or grants, loans or contracts provided to industry, universities and other performers outside of government.

THE AUSTRALIAN R&D TAX CONCESSION

The R&D tax concession in Australia is an income tax deduction equal to 125 per cent of eligible R&D expenditures.[4] An eligible taxpayer must be a company incorporated in Australia, a public trading trust, or an eligible company in partnership.

The definition of eligible R&D is based generally on the OECD definition of R&D.[5] In order to be eligible, R&D requires either the presence of an appreciable element of novelty or the resolution of scientific or technical uncertainty through a program of systematic and investigative and experimental activities. In addition, the work must be based on principles of physical, biological, chemical, medical, engineering or computer sciences. Furthermore, the R&D must satisfy certain 'Australian content' rules relating to key personnel and major items of plant and equipment, and the results of the R&D must be exploited on normal commercial terms and for the benefit of the Australian economy.

An annual minimum threshold of A$20,000 must generally be met for R&D spending to qualify for the tax concession.[6] Eligible R&D expenditures include current costs and capital expenditures on plant and machinery and pilot plant used exclusively for R&D.[7] R&D current expenditures are deductible at a rate of 125 per cent in the year incurred. R&D capital expenditures may be written off over three years on a straight-line basis. Expenditures for R&D carried on outside of Australia are also eligible if the amount of such expenditures does not exceed 10 per cent of the eligible expenditures for the associated R&D project as a whole.

THE AUSTRALIAN EVALUATION EXPERIENCE

The 1993 evaluation assessed the effectiveness of the Australian R&D tax concession in achieving its stated objectives, focusing in particular on the objectives of increasing investment in R&D by Australian companies, making Australian companies more innovative and internationally competitive, and establishing whether the assistance had a positive net contribution to economic welfare in Australia. The terms of reference for the evaluation also included analysing the intended and unintended impacts of the concession on the Australian industry and examining the overall relevance of program objectives within the context of industry, science and technology policy objectives.

Methodologies employed

The Bureau of Industry Economics used surveys and case studies to evaluate the impact of the Australian R&D tax concession on R&D spending in Australia, and on the innovativeness and competitiveness of firms in Australia. The evaluation also assessed the cost-effectiveness of the tax concession in terms of incremental R&D spending per dollar of tax revenue forgone and the value of the spillover benefits associated with the incremental R&D.

The survey/case study option was chosen, in part, because of problems with the quality and quantity of R&D data in Australian databases existing at the time of the evaluation. These data were inadequate to allow econometric analysis of, for example, the incrementality and spillover effects of the R&D tax concession. The surveys also allowed information to be obtained which helped to provide insights into the impact of the R&D tax concession on firms' innovativeness and competitiveness.

The rationale, objectives and findings underlying the surveys and case studies are discussed below. Some problems encountered in using each of these methodologies are also noted.

The surveys[8]

The surveys conducted by the Bureau of Industry Economics were intended to obtain the information necessary to address questions relating to the incrementality and cost-effectiveness of the R&D tax concession, and the impact of the R&D tax concession on the innovativeness and competitiveness of firms performing R&D in Australia. The incrementality (or cost-effectiveness) results were then combined with information obtained from a survey of foreign studies of spillover effects to estimate the potential spillover benefits accruing to the Australian economy as a result of the R&D tax concession.

The major source of data for the evaluation was a survey of R&D tax concession registrants conducted in mid-1992. This used a relatively large sample size (about 2,000 firms) so that the results could be considered meaningful and comprehensive. Overall, about 880 firms accounting for about 60 per cent of eligible R&D responded to the survey; this represented a response rate of about 44 per cent.

A second survey was conducted of R&D performers who were not registered for the R&D tax concession. These companies were selected from a database maintained by the Australian Bureau of Statistics which conducts an annual survey of R&D. Based on the results of this survey, about 620 firms formed a 'control group' whose R&D performance was compared with that of the recipients of the tax concession.

Incrementality and cost-effectiveness

The survey of recipients found that about 23 per cent of respondents felt the tax concession had been critical to their decision to proceed with at least one R&D project in the previous three years. There was little variation in this result by size of firm or taxable income status, although firms experiencing a higher rate of growth tended to indicate that they were more influenced by the tax concession. Projects influenced critically by the tax concession accounted for about 10 per cent of eligible R&D expenditures. Between 40 per cent and 50 per cent of respondents also indicated that the tax concession was a significant factor in their R&D work. Answers to another survey question indicated that eligible R&D expenditures would have been about 17 per cent lower if the tax concession rate had been reduced from 150 per cent to 100 per cent. This was particularly the case for small and medium firms, firms with fast growth in R&D, and Australian-owned firms. Taken together, these responses indicated that the amount of incremental R&D due to the tax concession was in the range of 10–17 per cent of eligible R&D expenditures. Drawing on this finding, and using information on the cost of the R&D tax concession, it was further determined that the concession induced between $0.60 to $1.00 of incremental R&D per dollar of tax revenue forgone.

A comparison with the results of the survey of non-recipients failed to shed additional light on the question of incrementality. While recipients of the tax concession displayed a higher rate of growth in R&D in some periods, only a small portion of the difference in R&D performance was able to be explained. The survey of non-recipients did reveal that about half of the firms were ineligible for the tax concession – 52 per cent of the ineligible non-recipients because of the threshold requirement and another 23 per cent because they were performing R&D on behalf of others. Of the non-recipients that were eligible for the tax concession, about 50 per cent indicated that they were in a tax loss situation while another 40 per cent cited administrative costs as being prohibitively high.

Innovativeness and competitiveness

The survey of tax concession recipients was also used to gain insights into the innovativeness and competitiveness of these R&D performers. Respondents indicated that in-house R&D was the most important source of new technology for them. This suggested a close link between R&D performance and innovation by these firms. However, since the ability to innovate (i.e., bring a new product or process to market) depends on a broad range of factors in addition to R&D, this linkage might be subject to considerable variability. Respondents provided information on the proportion of their sales in 1990/91 in each of four categories of products or processes: new, significantly improved, somewhat improved or unchanged. A weighted-average index of innovativeness was then constructed by assigning a declining set of weights to the proportion of sales in each of these categories. The results showed a clear positive correlation between the R&D intensity of firms (i.e. the ratio of eligible R&D expenditures to sales) and their degree of innovativeness.

The relationship between innovativeness and competitiveness was then examined by comparing:

- the performance of more innovative with less innovative tax concession recipients; and
- the performance of tax concession recipients with all manufacturing firms.

Performance was measured by considering the growth of sales, the ratio of operating profits to sales and the growth of profits relative to sales. The results revealed that less innovative tax concession recipients had lower rates of sales growth than more innovative recipients and, taken as a whole, tax concession recipients had higher rates of sales growth than all manufacturing firms over the period from 1988/89 to 1991/92. However, it is important to note that the direction of causality between sales growth and innovation could not be determined. The relationship between the profitability measures and innovativeness was found to be even less certain. The study also found that:

- smaller firms tended to suggest that innovativeness was more important in terms of broadening their product range, while larger firms felt that its benefits lay more in reducing costs and increasing production flexibility;
- innovations in products or processes were more important for competitiveness than innovations in management methods, marketing/distribution, financial management or workplace reform; and
- over half of the highly innovative firms ranked technology as very important for positioning in domestic markets, while less than 25 per cent of less innovative firms regarded technology in this way.

Spillover benefits

Gross spillover benefits from R&D in Australia were estimated based on a sensitivity analysis which used the incrementality (or cost-effectiveness) results for R&D spending from the survey of tax concession recipients together with estimates of spillover benefits for other countries (principally Canada) obtained from a literature survey. Drawing on the literature survey, the value of spillover benefits was assumed to be between \$0.66 and \$0.90 per dollar of incremental R&D.[9]

Net spillover benefits for Australia were then estimated by taking account of certain costs or leakages. These calculations assumed:

- a nil opportunity cost associated with reallocating resources from other sectors to conduct the incremental R&D generated by the 150 per cent tax concession;
- drawing on survey evidence, a 20 per cent leakage of spillover benefits to foreigners; and
- based on a literature survey, efficiency losses of between 15 per cent and 50 per cent of the tax revenues needed to finance the R&D tax concession.

The results of the analysis of net spillover benefits were mixed. Positive or negative results could be obtained by employing alternative combinations of the underlying parameter values and there was no strong prior reason for preferring one combination of parameters over another. However, using mid-point parameter values as the base case, the overall net spillover benefits were estimated to be 10 per cent of the cost of the R&D tax concession in terms of tax revenues forgone.

The case studies

The Bureau of Industry Economics also conducted case studies of fifteen innovating firms, sixteen of their innovations (two consumer products, two process innovations and twelve intermediate products or processes), and, where possible, some users of these innovations to try to obtain a better understanding of the nature, value and extent of spillovers from Australian research and development.[10] This work was intended not only to address the extent to which spillover benefits arising from innovations influenced by the tax concession flow to the Australian economy but also to help confirm the appropriateness of the foreign spillover estimates used in the evaluation of the R&D tax concession.

The case studies focused on three major aspects of spillovers:

1 Knowledge spillovers: spillovers of knowledge and know-how can benefit competitors or the innovative efforts of other R&D performers. This diffusion occurs through four major channels; namely, positive disclosure, interpersonal networks, labour mobility, and product availability/reverse engineering.
2 Increase in consumer surplus: firms and consumers can benefit from innovation through lower prices on existing products (process innovation) or new and improved products (product innovation).
3 Community benefits: society in general can also benefit from innovation in the form of improvements in environmental quality or public health and safety.

The innovating firms were all registered for the R&D tax concession; had responded to the 1992 survey; had reported that the tax concession had critically influenced their level of R&D; had recently introduced two or more significantly improved products or processes; and, had a high degree of R&D intensity. Interviews were conducted to gain an understanding of the nature of the innovation, pricing and marketing issues, the intellectual history of the innovation, the innovation process, how the firm attempted to preserve its intellectual property, the firm's perceptions of how the benefits of its innovation are shared with its customers, and the firm's perceptions of other non-appropriable benefits. After interviewing the firms, it was decided to pursue information on only one innovation per firm except in the case of one firm which had two innovations that were significantly different. This decision was based on a judgement as to the completeness of information for each innovation and whether the inclusion of a second innovation would provide added insight into the nature, value and extent of the spillover benefits.

Users of innovative products or processes were also interviewed either in person (44) or by telephone (12) to better understand the spillover benefits of the innovation. Major users, as identified by the innovating firms, were interviewed in the case of intermediate product or process innovations; distributors, in the case of consumer products. These interviews were seen as necessary for two reasons. First, the users usually received at least some spillover benefits. Second, the interviews helped to provide a more balanced view of the value of the innovation.

There were a number of difficulties with the interviews with the innovating firms and the users of innovations. In the case of innovating firms, for example, the information provided could not be independently verified – it was not possible, for reasons of confidentiality, to approach the firm's competitors to gain a fuller understanding of the market conditions and the extent of any knowledge spillovers. In addition, the case studies revealed that the survey responses of these firms were not always accurate. In the case of users, some intermediate innovations comprised only a small part of the final product so that their spillover benefits were difficult to isolate and estimate. As well, some users of innovations were either unable or unwilling to be interviewed.

In terms of their results, it is interesting that the case studies were unable to determine whether the tax concession was a critical factor influencing any of the sixteen innovations considered, while all but one firm indicated that the tax concession was critical to some part of the firm's R&D. As to the spillover benefits associated with the innovations, the three major aspects of spillovers were judged qualitatively using a value scale ranging from very low to very high, with the following results:

- Knowledge spillovers for the innovations were found to be very low in two cases, low in eleven cases and high in the remaining three cases. The size and structure of the Australian economy was identified as a major factor underlying the relatively low value of these spillovers, as was the number of domestic recipients of the knowledge and know-how.
- Consumer surplus spillovers for the innovations were found to be low in six cases, moderate in three cases, high in five cases and very high in the remaining two cases. These judgements were predicated on market structure, the lower costs or improved quality associated with the innovations, and the portion of sales that were domestic.
- Community spillovers for the innovations were found to be nil in three cases, low in eleven cases, and moderate in two cases. The spillovers were usually in the form of health benefits, although some involved environmental benefits as well.

Combining the results for the three categories, the total value of spillover benefits was found to be very low in one case, low in ten cases and moderate in five cases. While the study's authors indicated that these overall estimates are very rough, they felt that a qualitative estimate could at least provide a valuable preliminary indication of the spillover benefits to industrial innovation in Australia and indicate areas where further research would be required.

SOME LESSONS FROM THE AUSTRALIAN EXPERIENCE

The Australian experience provides several lessons with respect to using surveys and case studies to evaluate income tax assistance for R&D, and these may be relevant to Canadian evaluation work in this area. These include:

1 Information can be obtained through surveys and case studies which is not otherwise available in databases maintained for the purpose of administering tax incentives for R&D. For example, information can be obtained which allows insights into the impact of income tax assistance for R&D on firms' innovativeness and competitiveness, although the findings cannot be considered definitive.
2 The survey/case-study methodology can be effective in obtaining information on incremental spending attributable to R&D tax incentives and, consequently, on the cost-effectiveness of this tax assistance.
3 However, it is clear that the survey/case-study methodology must be carefully thought out in order to obtain accurate, unbiased and

meaningful information and the Australian experience shows that this is not always easily accomplished. For example, it could not be determined through the case studies that the R&D tax concession had indeed critically influenced any of the innovations considered, even though the firms' survey responses indicated that this was the case.

4 It may be useful to pose the same question in different ways to serve as a check on the accuracy of responses. This may be particularly important when there are no other external means by which answers can be corroborated. This approach was used to determine the incrementality of the R&D tax concession in the Australian study, and it was found that the responses to each question were different. As a result, a range of incrementality results was reported.

5 While answers to survey questions can provide information on the relationship between different elements within a study, care needs to be taken in interpreting the direction of causality between those elements. For example, while the Australian study noted a relationship between sales growth and the innovativeness of firms, it could not be determined whether higher sales growth was a consequence of, or a factor influencing, innovativeness.

6 Obtaining the cooperation of firms to participate in surveys or case studies can be difficult, but is essential if the validity of survey results is not to be put at risk. Large sample sizes are needed in the case of surveys in order to assure an adequate number of responses so that informed judgements can be made.

7 While, in some instances, the information obtained from the Australian surveys and case studies was not sufficient for addressing certain aspects of the evaluation work with a high degree of certainty, this does not necessarily cast doubt on the basic soundness of these methodologies for obtaining this information. These poor results may rather reflect, for example, the current industrial structure of the Australian economy. Furthermore, applying these same methodologies to larger R&D-performing countries (e.g. Canada) might provide more certain and meaningful results.

APPENDIX I: INCOME TAX INCENTIVES FOR R&D IN CANADA

The Canadian federal government provides substantial income tax assistance to businesses that perform SR&ED in Canada. This assistance, in the form of income tax deductions and investment tax credits, is a key component of the government's efforts to support and foster advancements in science and technology.[11]

In general, all of Canada's provincial and territorial governments follow federal income tax rules relating to the deductibility of current and capital expenditures in respect of SR&ED. However, six provinces – Manitoba, Newfoundland, New Brunswick, Nova Scotia, Ontario and Quebec – also offer various types of additional income tax incentives for R&D. The income tax support for R&D provided by the federal and provincial governments is widely recognized as one of the most favourable in the world (see Appendix II).

This Appendix reviews these income tax incentives. First, the federal tax policy objectives underlying the SR&ED tax incentives are outlined. Key elements of the current federal and provincial tax incentives for R&D are then described.

Federal tax policy objectives for assisting SR&ED

Federal income tax incentives for SR&ED assist the private sector in developing new products and processes, improving productivity, enhancing competitiveness and growth, and creating jobs for the benefit of all Canadians. Federal tax policy objectives in supporting SR&ED are to:

- encourage SR&ED to be performed in Canada by the private sector through broadly based support;
- assist small businesses to perform SR&ED;
- provide incentives that are, as much as possible, of immediate benefit;
- provide incentives that are as simple to understand and comply with and as certain in application as possible; and
- promote SR&ED that conforms to sound business practices.[12]

The federal SR&ED tax incentives

The federal government has provided income tax incentives for R&D since 1944 and there have been many changes over the years. Delivery mechanisms have included accelerated and incremental bonus deductions and investment tax credits, and have been designed to include both regional and small business considerations. The basic structure of the current federal income tax regime for SR&ED was put in place between 1983 and 1985, but has continued to evolve since then. Key elements of the current system of federal SR&ED tax incentives include (1) the definitions of SR&ED and allowable SR&ED expenditures, (2) income tax deductions, and (3) investment tax credits.[13]

The definition of SR&ED is consistent with the internationally accepted definition of RR&D used by the OECD. In particular, SR&ED is defined to be *systematic investigation or search carried out in a field of science or technology by means of experiment or analysis*. The three broad categories of work eligible are basic research, applied research, and experimental development. Basic research is *work undertaken for the advancement of scientific knowledge without a specific practical application in view*. Applied research is *work undertaken for the advancement of scientific knowledge with a specific practical application in view*. Experimental development is *work undertaken for the purposes of achieving technological advancement for the purposes of creating new, or improving existing, materials, devices, products or processes, including incremental improvements thereto*. The vast majority of the claims for the SR&ED tax incentives are for experimental development.

Certain support work is also eligible where such work is *commensurate with the needs, and directly in support, of* basic research, applied research or experimental development. To be eligible, the support work must be in respect of *engineering, design, operations research, mathematical analysis, computer programming, data collection, testing and psychological research*.

There is also certain work that is excluded from the definition of SR&ED – generally because it is not considered to be R&D in accordance with the internationally accepted OECD definition. Excluded work includes: *market research or sales promotion; quality control or routine testing of materials, devices, products or processes; research in the social sciences or the humanities; prospecting, exploring or drilling for, or producing, minerals, petroleum or natural gas; the commercial production of a new or improved material, device or product or the commercial use of a new or improved process; styles changes; or routine data collection*.

Current and capital expenditures in respect of SR&ED in Canada performed by, or on behalf of, a taxpayer and related to a business of the taxpayer, including a possible extension of that business, may be eligible for the SR&ED tax incentives. In addition, expenditures on equipment used primarily (more than 50 per cent) for SR&ED in Canada may earn a partial tax credit.

However, not all current and capital expenditures incurred for SR&ED in Canada are eligible for these tax incentives. For example, capital expenditures for the acquisition of land or buildings (other than a prescribed special purpose building), and current expenditures for related rental or leasehold payments are not allowable SR&ED expenditures. Also excluded are expenditures made to acquire rights in, or arising out of, SR&ED. Furthermore, some expenditures that are eligible for SR&ED income tax deductions are not eligible for SR&ED investment tax credits – for example, interest costs, legal and accounting fees, advertising or selling expenses. As well, while the costs of equipment used primarily for SR&ED in Canada may earn a partial tax credit, they are not eligible for an SR&ED tax deduction – instead, these expenditures are depreciable under the normal system of capital cost allowances.

In general, current expenditures that are eligible for the SR&ED tax incentives include:

- salaries or wages of employees directly engaged in SR&ED;[14]
- the cost of materials consumed in SR&ED;
- lease costs relating to machinery and equipment used all or substantially all (90 per cent or more) for SR&ED;
- eligible expenditures incurred by contractors performing SR&ED directly on behalf of the taxpayer;[15] and
- eligible expenditures incurred by certain third parties where the taxpayer is entitled to exploit the results of the SR&ED.[16]

In general, capital expenditures that are eligible for the SR&ED tax incentives consist of expenditures for machinery and equipment that is all or substantially all used or consumed in the prosecution of SR&ED in Canada.

Taxpayers have a choice in how they wish to treat overhead and administrative expenditures for a taxation year. They can use either the traditional method or the 'proxy method' for allocating these expenditures. The use of the proxy method is optional. However, once the choice is made, it is irrevocable for that taxation year.

Under the traditional method, overhead and administrative expenditures must be specifically identified and allocated in respect of SR&ED and may be eligible for both the SR&ED tax deduction and credits. Under the optional proxy method, these costs are deductible as ordinary overhead and administrative expenses and a notional amount is calculated which is eligible for the SR&ED tax credits. The notional amount for overhead and administrative costs is 65 per cent of salaries or wages (other than benefits, bonuses and unpaid amounts[17]) in respect of employees directly engaged in SR&ED – for example, the salaries of researchers carrying out experiments.[18] Salaries or wages of administrative staff that are providing a service to the SR&ED staff are not included as these amounts are captured as part of the overhead proxy (as are other types of overhead expenses).

Current expenditures in respect of SR&ED performed outside Canada by, or on behalf of, a taxpayer and related to the business of the taxpayer may also be eligible for an SR&ED tax deduction. In this case, the current expenditures must be either all or substantially all (90 per cent or more) attributable or directly attributable to the prosecution of SR&ED. Directly attributable current expenditures are defined by regulation to be costs of material consumed in SR&ED, salaries or wages of employees undertaking, supervising or supporting SR&ED, and other expenditures that are directly related to SR&ED and would not have been incurred in the absence of the SR&ED. The current expenditures may be incurred by contractors performing SR&ED directly on behalf of the taxpayer or, where the taxpayer is entitled to exploit the results of the SR&ED, by certain third parties.[19]

However, current expenditures incurred for SR&ED outside Canada are not eligible for the SR&ED tax credits. Capital expenditures in respect of SR&ED performed outside Canada are eligible for neither an SR&ED tax deduction nor the tax credits.

Both government and non-government assistance receivable by a taxpayer in a taxation year reduce the amount of expenditures available for the SR&ED tax incentives in that year. Government assistance is defined to include all forms of assistance from a public authority other than SR&ED tax credits. The amount of SR&ED tax credits claimed in a taxation year reduces the amount of eligible expenditures in the following taxation year. Non-government assistance includes any amount received by a taxpayer from any other person that can reasonably be considered an inducement, reimbursement, contribution, allowance or assistance.

Taxpayers are allowed to deduct fully eligible current and capital expenditures in respect of SR&ED incurred *in Canada* in the year. SR&ED expenditures that are not deducted in a year can be carried forward indefinitely. This is accomplished through the use of an SR&ED expenditure pool with an unlimited carry-forward period. SR&ED expenditures incurred in a year are added to the expenditure pool and can be deducted to the extent desired by the taxpayer. The pool balance remaining at the end of a year becomes the opening balance of the subsequent year.

There are two key differences between these income tax deductions for SR&ED expenditures and most other types of expenditures:

1 SR&ED capital expenditures can be fully deducted in the year incurred – capital expenditures are normally deductible over time through the capital cost allowance system; and
2 SR&ED current expenditures can be carried forward indefinitely – current expenditures are normally deductible only in the year incurred, and may create a non-capital loss which can generally be carried back three years or forward seven years.

Eligible current expenditures on SR&ED carried on *outside Canada* are fully deductible in calculating taxable income for a taxation year. However, this deduction is not the same as that provided for SR&ED in Canada. In particular, current expenditures on SR&ED performed outside Canada are not included in the SR&ED expenditure pool, cannot be carried forward and must be deducted in the year the expenditure is incurred. In general, capital expenditures on SR&ED incurred outside Canada are deductible under the system of capital cost allowances.

There are currently two rates of *investment tax credit* (ITC) for SR&ED in Canada: a general rate of 20 per cent and an enhanced rate of 35 per cent for certain Canadian-controlled private corporations (CCPCs); i.e. CCPCs with prior-year taxable income under $400,000 and prior-year taxable capital employed in Canada under $15 million. The amount of SR&ED expenditures that can earn ITCs at the enhanced rate is referred to as the expenditure limit.

The expenditure limit is generally $2 million for CCPCs with prior-year taxable income of $200,000 or less. This expenditure limit is reduced on the basis of the following two criteria. First, the expenditure limit is phased out for CCPCs with prior-year taxable income between $200,000 and $400,000. Specifically, for each dollar by which taxable income for the prior year exceeds $200,000, the SR&ED expenditure limit for the year is reduced by $10. In addition, the expenditure limit is phased out for CCPCs with prior-year taxable capital employed in Canada between $10 million and $15 million. Specifically, for every $10 by which taxable capital employed in Canada for the prior year exceeds $10 million, the SR&ED expenditure limit for the year is reduced by $4.

A partial ITC, equal to one-half of the normal credit, is also available for expenditures in respect of new equipment used primarily for SR&ED in Canada. This partial credit is earned in two instalments. The first instalment – one-half of the partial credit (i.e. one-quarter of the full credit) – is earned in the first taxation year that ends at least twelve months after acquisition of the equipment (i.e. the initial period). The second instalment is earned in the taxation year that ends at least twenty-four months after acquisition.[20]

ITCs may be deducted from federal taxes otherwise payable. Unused tax credits can be carried back three years (to the extent that they were not deductible in the year they were earned) or carried forward ten years. In addition, unincorporated businesses and certain CCPCs can obtain a refund of unused credits earned in a year. The general rate of refund is 40 per cent for both current and capital expenditures. However, current expenditures that earn SR&ED tax credits at the 35 per cent rate are fully refundable. Corporations can also assign expected refunds of SR&ED tax credits to lenders as security for bridge financing for their operations. Such assignments, however, are not binding on the Crown. Federal SR&ED tax credit and refundability rates are summarized in Table 5.1.

Provincial income tax incentives

All provinces generally follow federal rules in respect of the deductibility of current and capital expenditures on SR&ED. The provinces of Manitoba, New Brunswick, Newfoundland, Nova Scotia, Ontario and

Table 5.1 Federal SR&ED tax credit and refundability rates (%)

	Credit rate	*Refundability rate (current)*	*Refundability rate (capital)*
Unincorporated businesses	20	40	40
CCPCs with prior-year taxable income			
• of $200,000 or less:			
Expenditures up to expenditure limit*	35	100	40
Expenditures over expenditure limit	20	40	40
• between $200,000 and $400,000:			
Expenditures up to expenditure limit†	35	100	40
Expenditures over expenditure limit	20	0	0
CCPCs with prior-year taxable capital employed in Canada between $10 million and $15 million			
Expenditures up to expenditure limit‡	35	100	40
Expenditures over expenditure limit	20	0	0
All other corporations	20	0	0

Notes: * Expenditure limit is generally $2,000,000 per annum.
† Expenditure limit for CCPCs is phased out for prior-year taxable income between $200,000 and $400,000.
‡ Expenditure limit for CCPCs is phased out for prior-year taxable capital employed in Canada between $10 million and $15 million.

Quebec also offer various types of special income tax incentives for R&D conducted within their borders. These special provincial incentives, primarily investment tax credits (ITCs) and bonus deductions, are summarized in Table 5.2.

As mentioned previously, the federal SR&ED tax credits and deduction are based on qualifying expenditures net of government assistance. Under federal income tax rules, provincial ITCs are considered to be government assistance and reduce the amount of expenditures available for the federal SR&ED tax incentives in the year in which the provincial credits are earned.

Table 5.2 Summary of provincial R&D tax incentives

Province	*Tax incentive*	*Rate*
Manitoba	Non-refundable ITC	15%
New Brunswick	Non-refundable ITC	10%
Newfoundland	Fully refundable ITC	15%
Nova Scotia	Fully refundable ITC	15%
Ontario	Partially refundable ITC	10%: for certain CCPCs on SR&ED expenditures up to $2 million per year
	Super Allowance deduction	CCPCs: 35% up to base amount (average spending of previous three years); 52.5% on incremental R&D spending Non-CCPCs: 25% up to base amount (average spending of previous three years); 37.5% on incremental R&D spending
Quebec	Fully refundable tax credits for R&D salaries and eligible expenditures under various types of research contracts	40%: for small firms (assets under $25 million) on R&D salaries up to $2 million 40% to 20%: for medium firms (assets between $25 million and $50 million) on R&D salaries up to $2 million 20%: for large firms (assets over $50 million) and R&D salaries over $2 million 20% to 40% – for eligible expenditures under contract

APPENDIX II: AN INTERNATIONAL COMPARISON OF R&D TAX SUPPORT

A 1994 study by the Conference Board of Canada provides an international comparison of the income tax treatment for R&D investments in the G-7 countries and Australia. The countries were ranked by comparing the minimum benefit–cost ratio at which an R&D investment becomes profitable, given a country's income tax treatment for large manufacturing firms performing this work. Specifically, the minimum benefit–cost ratio is the present value of before-tax income necessary to cover the cost of an initial R&D investment and to pay the applicable income taxes. The lower the ratio, the greater the incentive for these firms to invest in R&D.[21]

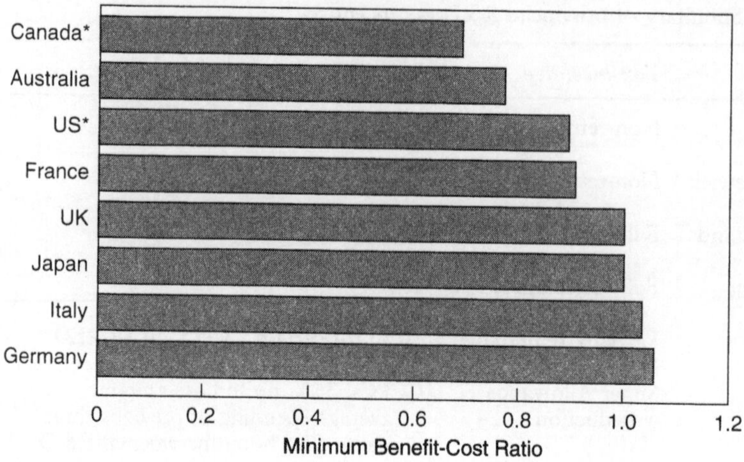

Figure 5.1 A comparison of R&D tax incentives in the G-7 and Australia (1993). * Canada: for an R&D investment located in Quebec; US: for an R&D investment located in California.

The study shows that, after taking account of both federal and provincial incentives, Canada has the most favourable income tax treatment for R&D investments among the G-7 and Australia. Australia, which provides a bonus income tax deduction (at the time, 150 per cent of eligible current and capital spending on R&D), is in second place. Germany, which does not offer special incentives for R&D, is in last place. Each of the other G-7 countries provides some form of income tax assistance for R&D. The results of this comparison are reproduced in Figure 5.1.

ACKNOWLEDGEMENTS

I wish to thank Paul Berg-Dick, Phil Feely, Bob Hamilton, Gérard Lalonde, Don MacDonald and Bill Toms for their helpful suggestions. The views expressed in this chapter are the responsibility of the author alone.

NOTES

1 See Bureau of Industry Economics (1993). At that time, the Australian Bureau of Industry Economics was an independent economic research agency attached to the Australian Department of Industry, Science and Technology.

2 Treasury Board of Canada (1992) provides guidelines for the conduct of program evaluations by the federal government. Under these guidelines, evaluations involve an empirically based analysis of the performance of programs in

achieving their objectives. Program performance, in turn, is defined as the continued relevance, success and cost-effectiveness of federal programs. *Relevance* is the extent to which a program continues to be consistent with departmental and government-wide priorities and to address realistically an actual need. *Success* is the extent to which a program is effective in meeting its objectives, within budget and without significant unwanted outcomes. *Cost-effectiveness* is the extent to which a program is the most appropriate and efficient means for achieving the objectives, relative to alternative design and delivery approaches and is delivered in the most cost-effective manner.

3 See OECD (1994), chapter 2, pp. 29–45. Chapter 1, pp. 18–21 discusses the distinction between R&D and other closely related activities that can be grouped, more broadly, under the headings of scientific and technological activities and scientific and technological innovation. Scientific and technological activities comprise systematic activities closely concerned with the generation, advancement, dissemination and application of scientific and technical knowledge in all fields of science and technology. These include such activities as R&D, scientific and technical education and training, and scientific and technological services. Scientific and technological innovation may be considered as the transformation of an idea into a new or improved product introduced on the market or a new or improved operational process used in industry or commerce. Innovations involve a series of scientific, technological, organizational, financial and commercial activities. R&D is only one of these activities, and may be carried out at different phases of the innovation process.

4 The maximum rate of deduction was reduced from 150 per cent to 125 per cent in the Australian budget of 20 August 1996. Prior to 24 July 1996, two or more Australian companies could also form a syndicate to contract out or undertake R&D in Australia. Syndicated R&D was complementary to the then 150 per cent R&D tax concession and the two had similar requirements concerning eligible R&D expenditures. In addition, a syndicate had to incur more than A$500,000 in total R&D expenditure to qualify for a rate of deduction in excess of 100 per cent. The policy intent of syndicated R&D was to allow a group of companies to undertake R&D projects that were beyond the resources of, or considered too risky for, a single company. One member of the syndicate was typically a smaller research company with tax losses that wished to undertake additional R&D on a pre-existing technology it possessed. Through syndication, this non-taxpaying research company could licence its pre-existing technology and transfer the tax losses relating to the pre-existing technology to the taxpaying corporate investors in the syndicate to obtain funding for the additional R&D. Each corporate investor in the syndicate could deduct its proportionate share of eligible R&D expenditures in calculating its taxable income. The rate of deduction for expenditures relating to the pre-existing technology was 100 per cent; the rate for additional R&D expenditures ranged between 100 per cent and 150 per cent, the higher rate being applicable to investments that were fully at risk. Claims could be made up to 13 months in advance of expenditure. Subject to grandfathering provisions for existing syndicates, the syndicated R&D program was terminated on 23 July 1996.

5 There are, however, differences between the two. For example, in Australia, certain activities such as routine testing and data collection, and market research and sales promotion are eligible if they are directly related to an 'eligible core activity'. In addition, computer software development is eligible if the software is developed for sale, but not eligible where it is developed solely for in-house use.

6 This threshold does not apply to contract payments made to a Registered Research Agency, i.e. an organization approved as being capable of undertaking contract R&D in one or more classes of activity for multiple clients. The use of these agencies allows taxpayers with smaller claims to access the 125 per cent R&D tax concession.

7 Capital expenditures on constructing or reconstructing buildings are generally deductible under the normal depreciation system over a period of 40 years on a straight-line basis. Interest and expenditures made to acquire pre-existing technology for the purposes of the taxpayer's own R&D are deductible at a rate of 100 per cent.

8 For further information, see Hawkins and Lattimore (1994) and Bureau of Industry Economics (1993).

9 See Dempster (1994).

10 See Bureau of Industry Economics (1994).

11 Total federal support for R&D equalled about $4.4 billion in 1995/96. Of this amount, $1.2 billion (based on 1992 data) was in the form of investment tax credits for SR&ED, while the remaining $3.2 billion consisted of various forms of non-tax assistance for R&D. Total federal support in respect of the broader category of scientific and technological activities (see note 3) equalled about $6.7 billion in 1995/96. Within this broader category, non-tax funding of scientific and technological activities was $5.5 billion of which $2.3 billion was for scientific activities related to R&D and $3.2 billion was for R&D *per se*. The SR&ED tax credits accounted for the remaining $1.2 billion.

12 See Department of Finance (1996), p. 174.

13 For a history of R&D incentives in Canada, see Hamilton (1993) and Clark *et al.* (1992).

14 There are special rules for salaries or wages paid to a 'specified employee' – a person who does not deal at arm's length with their employer or who has a significant interest (i.e. 10 per cent or more) in the shares of their employer. These rules place a ceiling on the amount of salaries or wages paid to these employees that are eligible for the SR&ED tax incentives. Salaries or wages of specified employees directly engaged in SR&ED are limited to a maximum of five times the year's maximum pensionable earnings for purposes of the Canada Pension Plan and exclude any remuneration based on profits or bonuses.

15 Where SR&ED is performed under contract between non-arm's length parties, expenditures eligible for the SR&ED tax credits are restricted to those incurred by the SR&ED performer. The performer can transfer these expenditures to the payor up to a maximum of the contract amount. The contract payment itself is not an eligible expenditure for tax credit purposes and does not reduce the eligible expenditures of the performer. In addition, where goods or services for SR&ED are purchased by an R&D performer from a person with whom the performer does not deal at arm's length, expenditures eligible for SR&ED tax credits are limited to the cost to the non-arm's length person of providing the goods or services.

16 Eligible third parties are corporations resident in Canada, including tax exempt non-profit SR&ED corporations and approved associations, universities, colleges, research institutes and organizations. In addition, tax exempt non-profit SR&ED corporations resident in Canada are eligible third parties if the taxpayer is a corporation and the SR&ED is basic or applied research that relates to other SR&ED being undertaken by the taxpayer and that has the technological potential for application to other unrelated businesses.

17 Current expenditures that are not paid within 180 days of year-end are deemed to have been incurred for SR&ED tax credit purposes in the year the amount is paid.

18 For purposes of the overhead proxy, salaries or wages of specified employees (see note 5) are limited to a maximum of two and one-half times the year's maximum pensionable earnings for purposes of the Canada Pension Plan and exclude any remuneration based on profits or bonuses.

19 These third parties are approved associations, universities, colleges, research institutes or other similar institutions which undertake SR&ED outside Canada.

20 These credits are only available to equipment that is used primarily in SR&ED during the initial period, i.e. the time between acquisition and the end of the first taxation year that is at least twelve months after acquisition. Equipment not used primarily in SR&ED during the initial period would never be eligible for partial tax credits.

21 Warda (1994). The report includes a description of the income tax regimes for R&D in each of the countries and further information on the methodology employed.

BIBLIOGRAPHY

Bureau of Industry Economics, *R&D, Innovation and Competitiveness: An Evaluation of the Research and Development Tax Concession* (Research Report 50) (Canberra: Australian Government Publishing Service, 1993).

Bureau of Industry Economics, *Beyond the Innovator: Spillovers from Australian Industrial R&D* (Occasional paper 16) (Canberra: Australian Government Publishing Service, 1994).

Clark, W. Steven, Goodchild, Gerry, Hamilton, Robert and Toms, William, 'Canada's R&D Tax Incentives: Recent Developments', *1992 Conference Report: Report of Proceedings of the Forty-fourth Tax Conference* (Toronto: Canadian Tax Foundation, November 1992).

Dempster, Peter, 'Empirical Estimates of External Returns to Business Expenditure on R&D: An Introduction to the Literature', *Working Paper No. 91* (Ottawa: Bureau of Industry Economics, February 1994).

Department of Finance Canada, *Budget Plan* (Ottawa: Department of Finance, 6 March 1996).

Hamilton, Robert, 'Tax Incentives and Innovation: The Canadian Treatment of R&D', *Canada–United States Law Journal*, Volume 19, April 1993.

Hawkins, Robert and Lattimore, Ralph, 'The Australian Experience in Evaluating the R&D Tax Concession', Paper Presented to the Conference on Government Recipes for Industrial Innovation, Vancouver, 20 and 21 October 1994.

OECD, *Frascati Manual 1993 – The Measurement of Scientific and Technological Activities: Proposed Standard Practice for Surveys of Research and Experimental Development* (5th edn) (Paris: OECD, 1994).

Treasury Board of Canada, *Evaluation and Audit Manual for the Government of Canada* (Ottawa: Treasury Board of Canada, January 1992).

Warda, Jacek, *Canadian R&D Tax Treatment: An International Comparison* (Report 125–94) (Ottawa: The Conference Board of Canada, 1994).

Part III

BANK FINANCING OF TECHNOLOGY DEVELOPMENT

Lending to small to mid-sized enterprises

6

THE BANKS AND INNOVATIVE ENTERPRISE

Opportunities and constraints

Jeffrey G. MacIntosh

INTRODUCTION

Much has been written of late about the 'new economy'. In the new economy, knowledge and skill-based work is the locus of economic activity, displacing both the resource sector and low-skill, labour-intensive manufacturing. Much has also been written about what the banks can and cannot do in financing firms that are participants in the new economy. This chapter addresses the role of the banks in financing innovative enterprise. The message is straightforward: banks are low-risk lenders. There are severe constraints on the extent to which banks can finance businesses lacking either collateralizable assets or receivables. While (as Paul Toriel makes clear in his chapter) all of the major Canadian banks have taken great strides in lending to the innovative (or 'high tech') sector, it seems doubtful that the banks will ever be in a position to lend to small, start-up enterprises lacking a proven product and a healthy record of sales (and hence receivables to finance interest payments and to provide collateral for the loan).

Some of the factors that account for the difficulty that young high tech firms encounter in securing loans are explored in the following section.

THE ECONOMICS OF LENDING TO SMALL HIGH TECH FIRMS

The absence of 'bankable'[1] assets

Lack of collateral

The availability of collateral is a vital condition for securing bank[2] credit in Canada, a result that has been replicated in many studies.[3] Acceptable collateral falls into two categories. The first is tangible assets like equipment and machinery, inventory, vehicles, land and buildings, and securities.[4] The second is accounts receivable.[5] High tech firms tend to be collateral-poor. For one thing, high tech firms are knowledge based, and thus lack tangible or fixed assets for collateralizing bank loans.[6] For another, the lag from firm creation to the availability of accounts receivable is longer for many high tech firms than traditional firms, given the long product development times normally associated with high tech products.[7]

Transaction specificity of assets

Assets that possess the property of transaction specificity are much more valuable in their current use than in any alternative use. For example, a biotechnology firm might design a machine to artificially manufacture blood products that can be used to replace the various components of human blood. If it turns out that the products are harmful to the human body, the machine, which might have cost millions to construct, may be worth little more than the salvage cost of its constituent parts.

Transaction specificity impairs the value of property as collateral.[8] Should a lender take security over the machine described above in order to secure its loan, for example, it is not likely to realize very much on its security should it be forced to seize the machine in the event of default.

High tech firms tend to possess transaction specific assets. Computer software, for example, tends to have a low resale value on insolvency.[9] Indeed, in a study by Wynant and Hatch, the *average* recovery on the seizure of small firm assets was only 64 per cent of the bank's outstanding loan and unpaid interest, even before legal and other wind-up costs.[10] Recoveries on high tech assets are likely to be lower.[11] Lenders will thus be reluctant to accept such assets as security for a loan.[12] The condition of transaction specificity thus disfavours the use of debt in the capital structure of high tech firms.[13]

Systematic and unsystematic risk

Systematic risk

A primary component of systematic (or non-diversifiable) risk is financial risk. The primary determinant of financial risk is the degree to which the firm has levered its capital structure (i.e. capitalized with debt rather than equity). Significant debt capitalization will result in substantial interest burden. This burden must be discharged whatever the economic environment the firm finds itself in. Moreover, the consequences of failing to pay interest can be severe; the creditors may put the firm into receivership or bankruptcy proceedings. Since bankruptcy is costly to the firm, higher levels of debt create greater financial risk.

However, the issue in this section is precisely the question of how much banks are willing to leverage small high tech firms. Thus, if high systematic risk is to be regarded as a deterrent to bank lending, it must be for some reason other than the lending itself.[14]

Another significant determinant of systematic risk is the degree of operating leverage: that is, the ratio of fixed to variable expenses.[15] High operating leverage impairs the ability of a firm to cut expenses when the economy turns down, resulting in a more significant reduction of the bottom line than for firms with low operating leverage.

Much of a small high tech firm's early stage financing will be spent on research and development (R&D), product development, and market research.[16] It could be argued that these expenses are highly variable; since the firm is producing no product, large variations in these variables can occur without affecting production.

This overlooks the fact, however, that while changes in these variables are easy to make, such changes will result either in abandonment or significant alteration of the project under development. A significant reduction in the R&D budget, for example, may effectively amount to a decision to abandon a prospective product and undertake (cheaper) research on an alternative product. For this reason, there is good reason to regard R&D, product development, and marketing expenditures as relatively fixed. On this view, young high tech firms can be characterized as exhibiting high operating leverage.

Thus, there is some reason to believe that high tech SMEs are likely to be characterized by relatively high systematic risk. Because banks are low risk lenders, this will tend to discourage bank lending.

Unsystematic risk

There are many factors that contribute to a high level of unsystematic risk. Not the least of these is that in the early stages of financing, high tech firms will be developing new products that are unproven and untested, and for which the market may be entirely speculative. Moreover, high tech products often involve lengthy product development and a short sales life.[17] There is also a substantial risk that before the investment can be brought to fruition, the technology will be rendered obsolete. Thus, the company's success is far from assured.

The quality of the management of small technology firms also varies widely. The absence of significant operating history and/or information about the quality and character of the managers makes it difficult for creditors and investors to evaluate managerial quality *ex ante*. This adds to the unsystematic risk faced by creditors and equity holders.

In theory, lenders holding large portfolios of loans should be unconcerned about unsystematic risk. This risk if fully diversifiable. This theory, however, does not appear to hold in practice. Banks, by their very nature, are highly leveraged; approximately 92 per cent of assets are in the form of deposits.[18] Thus, banks are themselves subject to considerable financial (systematic) risk. This does not, in theory, mean that banks should avoid lending to clients with high unsystematic risk. Increases in fully diversifiable unsystematic risk do nothing to augment a bank's nondiversifiable risk. However, in practice it is difficult for account managers to distinguish between systematic and unsystematic risk factors. Indeed, it is probable that many do not understand the difference. Thus, it is rational for banks to instruct loan managers to avoid loans to firms with high risk, without distinction as to the type of risk.

Wynant and Hatch document Canadian banks' conservative lending practices. The banks aim at bad debt losses on the order of 0.5 per cent of outstanding loans; the target for non-accrual loans (i.e. loans on which interest is not being paid) is on the order of 1 per cent.[19] Moreover, in response to calls for more lending to the small business sector, Canadian banks have consistently and publicly adopted the posture that they are 'low risk' lenders, and are not in the business of financing high risk enterprises.[20] Lending guidelines for account managers[21] do not in fact distinguish between systematic and unsystematic risks. Moreover, the training of loan officers is geared to inculcating conservative lending practices.[22]

In short, both theory and evidence suggest that not merely systematic, but unsystematic risks will be taken into account in evaluating a particular lending prospect. Firms with high unsystematic risk are thus less likely to get bank loans.

The centrality of the entrepreneur

The most important asset of a small technology firm will be the innovative concept that it seeks to develop. This asset is typically inseparable from the firm's key personnel.[23] The key personnel will generally be the inventors or innovators who have hatched the idea that forms the inspiration for the enterprise. Should these personnel leave the firm, there will often be little point in attempting to keep it alive.

The centrality of the entrepreneur emphasizes the problem that lenders face in securing the debt of a high tech enterprise. Taking security over the assets of the firm is likely to yield a low recovery upon insolvency; once the human and non-human capital have been separated, the latter is likely to have little value.[24]

Evidence of the centrality of the entrepreneur arises in actual lending practices. Granting of bank credit to a small business will often be contingent on the continued involvement of the entrepreneur(s) and/or key shareholders. Failure to abide by such restrictions can result in the calling of the loan.

Lending costs per dollar of loan

Lenders incur a variety of costs in investigating loan applicants, negotiating the deal, and monitoring the loans. Many of these costs are relatively fixed over a wide range of loans, from small to large. The result is that the costs of lending to a small firm, per dollar invested, are greater than the comparable costs for larger firms. This results in narrower profit margins on small loans.

These costs include the following:

Selection costs

Lenders must invest time and effort in selecting their borrowers. The 'due diligence' required will not consume significantly less time for small loans than for large.[25] Indeed, it is often the case that the due diligence required to investigate a smaller firm will exceed that required for a larger firm. Smaller firms frequently do not have sophisticated accounting and information systems in place. It is thus more difficult for lenders to assemble the information they require.

The cost of negotiating and documenting the loan

The cost of negotiating and writing loan contracts is relatively fixed, leading to higher loan costs for small businesses.[26]

Monitoring costs

Lenders will wish to monitor their investments. The costs of monitoring typically do not increase in proportion to the size of the firm. Pettit and Singer, for example, suggest that 'charges for accounting and auditing systems are not proportional to the size of the firm'.[27] Similarly, it will likely take about the same time to read and digest the financial statements of a small company as a large one. It will also take a similar amount of time to inspect the firm's facilities and to perform other monitoring tasks. Thus, even assuming that agency costs are no greater for smaller than for larger firms (an assumption that is challenged below), and that monitoring costs are a positive function of firm size, monitoring costs per unit of loan will be higher for smaller firms.[28]

The cost of realizing on collateral

The cost associated with realizing on loan security by seizing firm assets will be relatively fixed. As firm size increases, these fixed costs are spread over a larger quantity of assets. In short, it seems likely that there will be economies of scale associated with the seizure of assets. This means that the cost of realizing upon small firm assets are likely to be greater per dollar of collateral and per dollar of investment than for large firms.

Reorganization costs

Should it be necessary to reorganize the firm to avoid insolvency, the costs of reorganization will be an increasing function of the number and nature of the financial claims on the firm, and the depth of the firm's physical and human capital. None the less, some costs of reorganization are relatively fixed. Thus, the costs of reorganization will consume a larger fraction of the assets of a small firm than of a large one.[29]

Summary of lending costs

It will be noted that some of the above costs (like borrower selection costs) are invariably incurred by all lenders. Others, like reorganization and seizure costs, are merely prospective costs with an associated probability function. The latter, however, have no less important an affect on the per dollar cost of lending than the former. Lenders will rationally anticipate, for example, that reorganization or seizure of assets *may* be necessary, and will factor these costs into their estimate of the expected return from the loan. All of the enumerated factors thus lead to higher lender costs per dollar of loan for small firms.[30]

The agency costs of debt

Debt holders are exposed to the risk that the managers, who in law are the representatives of the shareholders,[31] will take actions that have the effect of redistributing wealth from debtholders to shareholders. A lender contemplating a loan to a high tech firm will clearly be interested in knowing whether the agency costs of debt are likely to be higher or lower for small high tech firms than for other commercial clients. Higher agency costs will tend to reduce loan profitability, while lower agency costs will tend to increase it. In this section I argue that the agency costs of debt are likely to be higher for smaller firms.

A typology of the agency costs of debt

There are four well-known types of agency costs of debt.[32] I have added two others.

Risk shifting

Equity holders can transfer wealth from debtholders by increasing the risk of the firm's investments. This risk is likely to be higher for lenders to small firms. Typically, in such firms, ownership is concentrated in the hands of the entrepreneur/managers. This concentration increases the incentive of the managers to engage in risk shifting; as insider equity ownership rises, insiders capture a larger and larger share of any wealth that is redistributed as a consequence.[33] Moreover, in small firms insiders have an enhanced ability to engage in risk shifting, precisely because their equity interest gives them the powers of control needed to consumate risk-shifting events.[34]

At least one factor, however, will operate to mitigate the incentive of entrepreneur/managers to engage in risk shifting. Managers are commonly assumed to be underdiversified, given that a disproportionate share of a manager's wealth is tied to the fortunes of the firm, via managerial share-holdings, stock options and other incentive arrangements, and salary.[35]

An underdiversified agent is exposed not merely to the systematic risk of the enterprise but to the unsystematic risk as well. This is simply a function of the fact that diversification substantially reduces or eliminates unsystematic risk. Risk is a cost to risk averse agents; managers who undertake behaviour that increases systematic risk impose an added risk cost on themselves. Thus, underdiversified agents will tend to avoid behaviours that increase *either* systematic *or* unsystematic risk, even to the point of failing to adopt the highest net present value projects available to the firm, if these strategies are characterized by high risk.[36]

The problem of underdiversification will be particularly acute for entrepreneur/managers of small firms. Such managers will typically have a larger proportion of their wealth tied up in the firm than managers of large companies. Moreover, entrepreneur/managers of high tech firms will tend to have more transaction-specific human capital than managers of large firms, whose skills will generally be more fungible. High transaction-specificity will result in severe erosion of the manager's human capital should the firm fail.[37] This will naturally make the manager more reluctant to engage in risk-enhancing behaviour, since such behaviour will heighten the probability of firm failure and the consequent loss of a good deal of the manager's human capital.

Finally, the very fact that high tech firms are already highly risky will tend to limit the opportunity (and the inclination[38]) of the managers to increase risk in order to effect wealth transfers from creditors.

It should be noted, however, that entrepreneur/managers of small high tech firms have chosen to engage in a comparatively high risk undertaking. This self-selection conveys information about the risk preferences of such managers; on average, we would expect that they are less risk averse than managers of larger and less risky businesses. Some may even be risk preferring. The less the degree of risk aversion possessed by the managers (or the higher the degree of risk preference), the more will be the managerial inclination to engage in risk shifting.

Because of the conflicting forces at work, it is difficult to make theoretical predictions about whether the danger of risk shifting is greater or smaller for small high tech firms than for larger, non-high tech businesses (or indeed other small businesses).

Asset stripping

Entrepreneur/managers can shift wealth from debtholders to themselves by stripping assets out of the firm; for example, by paying excessive dividends or excessive managerial remuneration.[39] The same factors that give the managers of small firms the incentive and the means to engage in risk shifting give them the incentive and means to convert firm assets to their own use.[40]

The danger of asset stripping will generally be more serious for small firms,[41] and will be particularly serious for small technology firms. Larger, non-technology firms typically have a greater proportion of tangible assets than many small technology firms. This affords creditors enhanced protection against risk shifting; tangible assets are more difficult to move physically than the intangible assets often owned by small high tech firms. It is much easier, for example, to walk away with a computer program on a 3.5 inch disk than it is to walk away with a machine that is bolted to a factory floor. It is also easier for a lender to take security over the machine and trace it should it disappear. In addition, a large firm will generally have more managers and more employees. Asset stripping will be more difficult to effect without detection by some of these individuals.

That asset stripping is a particular problem for small firms is attested to by the fact that lenders almost always insist on personal guarantees when loaning money to small businesses.[42] A personal guarantee will reduce the incentive to take assets out of the corporation, since on default the lender can seize the entrepreneur's personal assets without having to prove that assets were siphoned out of the corporation.[43]

Underinvestment

Particularly as the firm approaches insolvency, the equity holders lose the incentive to adopt all positive net present value projects. The cost of such projects may be shouldered by the equity holders,[44] because most or all of the benefit of such projects accrues to the debtholders.[45] This problem is probably no more serious in small technology corporations than in large non-technology concerns.

Trumping debtholders' claims

Management can redistribute wealth from debtholders to equityholders by issuing debt claims equal or prior to the existing debt claims. Indeed, the issuance of subordinated debt claims may also redistribute wealth to the extent that it increases the burden of fixed interest charges and enhances the probability of a receivership or bankruptcy in which existing creditors will realize only a fraction of their claims.[46]

The comparatively large equity stake of small firm managers creates a greater incentive to engage in this type of wealth redistribution. This is simply because, as significant equity holders, the managers capture a large share of any redistributed wealth.

As indicated below, in small firm loans of $500,000 and over, this type of problem is often controlled through the use of restrictive covenants. One cannot simply assume, however, that these covenants will be equally effective in constraining opportunism on the part of both small and large firms. As discussed above, the cost of policing and enforcing such covenants will be higher, per dollar of loan, for small firms. To the extent that this results in a lower level of monitoring or enforcement, such covenants will be breached more frequently by managers of small firms.

The risk that very small firms will issue additional debt claims is even higher. The banks do not find it economical to negotiate, write, monitor and enforce restrictive covenants regulating or forbidding the assumption of additional debt for loans of less than $500,000. Without such covenants, lenders obviously take a greater risk that added debt claims will be issued.

This added risk of redistribution is no doubt one of the reasons why lenders to small firms typically require extensive collateral and a personal guarantee from the entrepreneur.

Shirking

One of the commonly acknowledged risks faced by outside equity holders is the danger that the managers will shirk; that is, that they will fail to work diligently and conscientiously on behalf of the firm. As residual claimants, equity holders stand last in priority with respect to both the earnings stream and assets on winding up. For obvious reasons, shirking will have a more substantial impact on residual claimants than on non-residual claimants. None the less, although less commonly acknowledged, it is clear that shirking will have an affect on the value of the debt claims.

How substantial will this impact be? If the value of the residual claim is significant, then a reduction in the value of the residual claim due to shirking will not greatly prejudice the likelihood that creditors will be paid. Hence, it will have a small affect on the value of the debt claims. However, as the size of the residue shrinks, further reductions obviously pose a greater threat to the ability of the firm to make good on its debt. Thus, the impact on the value of the debt claims will be greater. When the value of the residual claim becomes negative, further shirking is likely to have a greater adverse impact on the value of the debt claims than the equity claims.

It seems likely that shirking will be less of a problem for small technology firms than for other firms, for two reasons. One is that small firms are characterized by higher levels of managerial share ownership. The higher the manager's ownership stake, the greater is the proportion of the cost of shirking that will be borne by the manager. The second is that managers of high tech small firms will, by their very nature, tend to be hard-working entrepreneurs. It seems unlikely that they will have a significant propensity to engage in shirking.

Incompetence

While the firm's managers may have no predilection for shirking, they may well be incompetent in one or more respects. The likelihood of incompetence appears to be greater both for small firms generally, and small technology firms in particular.

The greatest source of incompetence in small firm managers is a lack of expertise in matters of financial management. Small firms are more likely to receive credit when at least one of the managers has financial management experience.[47] The absence of such expertise is a common cause of small firm failure.[48]

An absence of financial management skills is a particular problem for high tech businesses. Those launching high tech businesses are often good scientists but poor financial managers. Indeed, many scientists seeking funding for an innovative idea will be embarking upon a business career for the first time, and will thus have little or no business experience.

Another talent frequently lacking in high tech entrepreneurs is the ability to market the firm's product. The annals of entrepreneurship are littered with the corpses of enterprises which foundered for want of effective marketing, even though their products were clearly better than those of their competitors.

In short, lenders face a higher risk of incompetence with respect to small technology concerns than with respect to other firms.

The role of asymmetric information

Agency costs of debt are typically increased by the presence of asymmetric information.[49] This is illustrated by imagining a world in which monitoring costs are zero. In such a world, all agency costs are immediately detected. Hence, a contract can be written *ex ante* that will pass the full amount of the agency costs back to the unfaithful agents. The more we depart from this perfect world, the more likely it is that the managers can (and will) engage in undetected redistribution of wealth from creditors to equity holders.

High tech firms will almost certainly suffer from a higher degree of information asymmetry than non-technology tech firms. Information asymmetry is likely to be a function of two factors. The first is the supply of information. The second is the comparative ability of insiders (managers) and outsiders (lenders) to understand this information.

In respect of the first, the supply of information is frequently poor for small firms. Unlike their large firm counterparts, small firms will typically not have sophisticated information and accounting systems. For example, such firms are typically not required by law to be audited,[50] and many choose to forgo the expense of an annual audit. Hence, it is more expensive for outsiders to gather information pertinent to evaluating the riskiness of the firm. In addition, small firm owners often do not understand the importance of communicating information to lenders and will do so only reluctantly.[51]

The availability of information is also diminished given that smaller businesses are often young businesses. Such businesses lack a track record which might otherwise be used to evaluate the firm and its managers. Moreover, the quality of small firm managers exhibits high variance, and is difficult for outsiders to observe *ex ante*. Given the importance of the entrepreneur/manager to the fate of enterprise, it is thus not surprising that a survey of venture capitalists by the Premier's Council disclosed that '[e]arly stage venture capital investors often see themselves as investing in the entrepreneur and his potential rather than in a business'.[52] This 'management risk' is perhaps the most important result of the information asymmetry confronting lenders and investors.[53]

Aside from the difficulty of evaluating the competence of management, there are three other aspects of management risk. One is that outsiders cannot easily gauge managers' risk preferences. The degree of risk aversion or risk preference exhibited by the firm's managers will be an important attribute in the lender's estimate of the danger of risk shifting. Difficulty in evaluating managerial tastes for risk accentuates the overall management risk.[54]

Similarly, outsiders cannot easily observe the extent to which small business managers may wish to indulge non-pecuniary preferences in running their businesses. Some, for example, may prefer publicity or notoriety over profits. Indeed, non-pecuniary preferences (such as 'being one's own boss') are often said to be an important motivation for seeking self-employment. It is likely that the variance in preferences amongst small firm managers is greater than that found amongst managers of larger firms.

Finally, outsiders will not easily be able to observe the managers' work/leisure preferences, and so will find it difficult to predict the extent to which managers will engage in shirking.

Thus, management risk is likely to be high for small firms. It is not surprising, therefore, that loan turndowns are much more likely when the managers have little financial experience and/or the firm has little track record.[55]

Information asymmetry results not only from a shortage of information but also from the inability of those examining the information to evaluate that information properly (whether due to time constraints or lack of expertise). There is good evidence that bank managers are poor monitors of small businesses. Wynant and Hatch's study of the relationship between Canadian banks and small businesses disclosed that small business account managers typically manage between 75 and 150 accounts.[56] For obvious reasons, this prevents the account manager from spending a great deal of time on any one account. The problem is exacerbated by the fact that bank managers experience frequent internal reassignments. Wynant and Hatch found that '[o]ver 60% of small business clients have experienced at least one change in account manager during the past three years and 22% have dealt with three or more bankers'.[57] A study done for the Canadian Federation of Independent Business puts these figures at 76 per cent and 27 per cent respectively.[58] Such rapid turnover does not allow managers much time to become well acquainted with their business clients.[59]

Moreover, account managers often do not have the training or background to be able to conduct a sophisticated analysis of their small business clients. Historically, many Canadian account managers have entered banking straight from high school, and have worked their way up through various bank positions into commercial lending.[60] While this pattern is changing, and the banks are now hiring people with university training and promoting them (with special training) directly into commercial lending, it is still the case that few account managers have any business experience.[61]

The inability of bank managers to evaluate their clients will be particularly acute in relation to high tech businesses. Even in the venture capital arena, where participants bring a wealth of business experience with them, successful venture capitalists are increasingly focusing their attention on small areas of the high tech spectrum.[62] It seems unlikely that bank managers are able to understand very much about the businesses of their high tech clients.

Finally, because of the fixed component of monitoring cost, intensive monitoring is not cost-justified for many small business clients.[63]

Evidence of the inability of bank managers to engage in much useful monitoring of small businesses arises in the fact that small business managers typically feel that they get little useful business or financial advice from their account managers.[64] This inability also accounts for the heavy emphasis that banks put on the availability of collateral in determining creditworthiness. Wynant and Hatch suggest that:

> many bankers admit that they often can achieve only a limited understanding of the management team, the business and its true risks. As a result, collateral security provides an additional level of comfort to an account manager by ensuring that the overall level of credit risk is within acceptable bounds.[65]

Summary

Of the six types of agency costs examined, three (asset stripping, trumping debtholders' claims, and managerial incompetence) are likely to be more serious for such firms. One (shirking) is less likely to be a serious problem. One (underinvestment) is not likely to be different for small technology firms than for other firms, and one (risk shifting) is difficult to categorize on the basis of theory, owing to conflicting forces at work.

Information asymmetry enhances agency costs. Information asymmetry is clearly greater between small technology firms and their bankers than for larger and non-technology concerns.

Thus, in net, the agency costs of debt are likely to be higher for small technology firms than for larger firms and non-technology firms.

Market constraints on the agency costs of debt in small corporations

Control market

One of the primary constraints on agency costs in public corporations is the corporate control market.[66] When a firm whose shares are publicly traded performs poorly, it may be subject to a 'hostile' takeover bid (i.e. one opposed by management). The result of such a bid will often be that some or all of the managers are replaced by the acquiror.

Because private firms lack publicly traded claims and have a high concentration of equity ownership claims, they are less subject to control market correction of agency costs than public companies. Indeed, if there is a controlling shareholder, or coalition of controlling shareholders, a hostile acquisition is impossible.

This is not to say there is *no* control market, however, for the shares of private firms. Suppose that an acquiror identifies a private firm as an attractive acquisition candidate, and makes an offer to the shareholders at a price in excess of the firm's current value. Should the shareholders refuse the offer, they incur an opportunity cost equal to the difference between the offer and the value of the firm under existing stewardship. This opportunity cost may prompt controlling shareholders to participate in a consensual control transaction.

None the less, no offer will be forthcoming unless the offeror has reasonably good information about the value of the firm. Reliable information about the value of private firms is almost always hard to come by. By contrast, public companies are subject to disclosure obligations and often closely followed by analysts and investors.

Further, entrepreneur/managers of private firms may pursue a non-pecuniary agenda in seeking to maintain control of the firm. The existence of a non-pecuniary agenda will lower the opportunity cost of refusing to participate in a consensual transfer of control, even one which appears to an objective outsider to be a good deal. Non-pecuniary objectives can be observed in public companies as well as private. For example, Morck and Strangeland find that companies with founding entrepreneurs at the helm tend to be less profitable than widely held firms.[67] One explanation is that entrepreneur-founders realize significant non-pecuniary income by retaining control of the firms that they have founded.[68]

It is thus clear that the market for corporate control will function much less effectively in disciplining managers of private firms.

Managerial labour market

Managerial labour markets operate to constrain managerial opportunism in two ways. First, managers within the firm compete for promotion, and this creates an incentive for underlings to perform their jobs effectively. Second, if a manager performs poorly, this will increase the probability of being terminated. It will also lower the manager's prospects of securing an equally or more advantageous job elsewhere.[69]

In larger firms, internal labour markets are likely to operate effectively. Large organizations are typically characterized by many hierarchical layers, and the opportunities for internal promotion are great. This provides junior managers (or prospective managers) with a good incentive to excel.

By contrast, the internal labour market in small firms is typically rudimentary, if not non-existent. The number of managers, and layers in the managerial hierarchy, will be small; small firm command structures are often more horizontal than vertical. Opportunities for promotion will be limited.

In addition, the external labour market is likely to operate less effectively, because it will be difficult for outsiders to garner useful information about the performance of a small firm manager. There will be fewer people in a position to evaluate the work of the manager. Moreover, since managerial skill varies substantially in small companies, there will often be suspicion that those providing evaluations are not themselves good managers (and therefore not good evaluators).

While small firm managers will typically be a larger factor in the success or failure of the firm than large firm managers,[70] even the failure of a small firm may be a weak signal of managerial performance; many exogenous factors can be pivotal in causing small firm failure.[71]

Thus, managerial markets will provide a less effective constraint against managerial overreaching in small than in large firms.

Capital market

Managerial quality will affect the firm's cost of capital. This is commonly said to be a constraint on the behaviour of the managers of publicly held firms. Indeed, superior quality executives can lever their advantage over less effective managers by paying high dividends. It has been argued that this will increase firm value by heightening the firm's need to tap public markets for capital, which in turn will result in greater heightened market oversight and more discipline of managers.[72] Alternatively, better quality managers can signal their quality by causing the firm to take on a large amount of debt (thereby subjecting themselves to the greater discipline of having to meet substantial interest payments).[73]

Unlike the other market constraints explored here, the cost of capital is likely to be a more potent constraint for private than for public firms. Capital markets for small firms are characterized by all of the infirmities outlined earlier. In particular, outsiders will have poor information about the quality of the firm and its managers, and the costs of committing capital to the firm will be higher per dollar of capital than for larger firms. Because of these problems, many small businesses – even those 'deserving' of outside investment on a risk adjusted basis (assuming perfect information) – will have a difficult time accessing outside sources of capital.[74]

Small business owners do in fact spend a lot of time worrying about gaining access to capital, typically identifying a lack of access to both equity and debt capital as the biggest problem they face.[75] With many smaller firms on shoestring budgets, failure to succeed in the next round of financing will spell doom. The capital market constraint thus appears to be more binding for small firms than for larger firms.

Product market

If the firm cannot sell its products, it will go bankrupt, and the managers will lose their jobs; this motivates the managers to produce profits. Like the capital market constraint, this constraint is likely to be more binding for small companies, and particularly for small high tech companies. The reasons have already been explored; product market risk is high and the failure of a product will frequently mean a quick death for the firm. Further, as pointed out earlier, managers of private firms stand to lose more in bankruptcy than the managers of public firms, to the extent that they have a higher degree of firm-specific human capital that will be lost or compromised on firm bankruptcy.

Sum of market controls

It is difficult to predict whether market factors will have a lesser or greater effect on managerial agency costs for small, private firms than for large public firms. Two market constraints (control market and managerial market) are likely to be less binding, and two (capital market and product market) are likely to be more binding.

Lender responses to the high risk of high tech enterprises

Lenders adjust for the heightened costs and risks of lending to small, high tech enterprises in a number of ways. These include the following:

Collateralizing the loan

Aside from personal guarantees, the main response of Canadian banks to problems of shareholder opportunism has been to secure loans against tangible collateral with significant resale value. The type of collateral that the banks will accept includes accounts receivable, equipment, machinery, inventory, vehicles, land and buildings.[76] The banks almost never make loans against intangible assets that can be easily spirited off by the shareholders and/or whose resale value is poor.[77] As Wynant and Hatch report, '[a] great majority of bankers that we interviewed saw strong collateral backing as a "must" condition for small business loans'.[78]

Securing personal guarantees from the entrepreneur/manager(s)

Obtaining a personal guarantee (and/or securing personal assets) reduces the incentive to engage in asset stripping. If there is a default, the lender can access the personal assets of the entrepreneur/manager without the need to prove that assets were improperly appropriated. Guarantees are also a means of increasing the entrepreneur's commitment to the enterprise.[79]

In Canada, personal guarantees are routinely required by bank lenders. Wynant and Hatch, for example, found that personal guarantees were required in connection with 87 per cent of small business loans, although pledges of personal assets were required only one-third of the time.[80]

Manager screens

One of the problems faced by lenders is the absence of managerial business expertise and the risk of managerial incompetence, particularly in the area of financial management. Banks typically address this risk by making credit lesş available to firms without track records, and to firms lacking a manager with financial management skills.

Debt covenants

Debt covenants can be written to constrain various aspects of manager/shareholder opportunism. Covenants can be written, for example, to require the maintenance of certain financial ratios, minimum working capital, lender approval of any disposition of a major asset, restricting the right to pay dividends, etc.[81]

Generally speaking, in loans of under $200,000, Canadian lenders typically insist on few or no covenant restrictions. Where there are covenants, they typically relate to matters such as subordination of shareholder claims, life insurance on key principals, fire insurance on the company's premises, and reporting of accounts receivable and inventory.[82] They typically do not touch matters of corporate governance or the ability to enter into major transactions.[83] Although the risk of opportunism may be higher for the smallest firms, covenants are not a cost-effective means for constraining such opportunism. This is true for two reasons. First, given the fixed costs of negotiating and writing debt covenants, it is proportionally more expensive to do so for smaller firms. Second, and perhaps more importantly, restrictive covenants are meaningful only where there is a monitoring apparatus in place to police for covenant violations and to secure enforcement of the covenants. Because margins on small loans are thin, engaging in more than rudimentary monitoring is usually not cost-justified. Thus, there is little to be gained from inserting restrictive covenants into the loan agreement.

An added reason why covenant protection tends not to be cost-justified for small firm loans is that the opportunity cost of writing covenants will be comparatively great for smaller – and particularly for high tech – firms.[84] Such firms exist in fast moving and highly fluid markets, and the ability to make quick changes in managerial strategy and asset mix may be vital to maintaining a competitive edge. Even if it were economical to enforce debt covenants, doing so might constrain management's ability to respond quickly to market changes or to the introduction of new technologies by competitors, either because covenants prevent management from undertaking particular actions, or simply slow down management's ability to take such actions.

Thus, in making loan decisions, loan officers will focus primarily on available collateral, the willingness of the principals to give personal guarantees, and the character and managerial experience of the firm's principals.

In loans between $200,000 and $500,000, it becomes economical to make use of restrictive covenants, although these are generally restricted to simple financial ratios (like maintenance of specified debt/equity and working capital ratios) and/or a requirement for maintenance of working capital.[85]

In loans over $500,000, additional covenant protection may be included. Anecdotal evidence suggests that the most common form of additional protection is a covenant indicating that if the principal shareholders of the firm change, the loan will become due and payable. Other provisions might include restrictions on the maximum expenditure the firm can make without the approval of the lender; the issuance of further debt; the disposition of major corporate assets; and the payment of dividends. In addition, there may be a 'significant material changes' clause under which the lender can accelerate its debt on the occurrence of a significant material change in the firm. The last type of clause effectively replaces many detailed restrictions with a single restriction of very general import. It is rare to see all of these clauses in a single loan document, and in general, the degree of covenant protection even for small firm loans in excess of $500,000 will be fairly rudimentary.

Shortening the term of the loan

By shortening the term of the loan, managerial opportunities to engage in opportunistic wealth redistribution are commensurately attenuated. The lender is also given more frequent opportunities to re-evaluate the loan and penalize cheaters by adjusting the interest rate, the terms of the loan, or simply by refusing further credit.

It is frequently stated that there is a shortage of both long term debt and equity in Canada.[86] There is evidence to the contrary; Wynant and Hatch found that the average term of a small business term loan was 55 months. It should be noted, however, that just over 80 per cent of the loans were for periods of less than 60 months, and the median term was 48 months.[87] The majority of long term lending was against land and buildings.[88]

Looking only at term loans to gain an appreciation of the average duration of bank loans, however, is misleading, because it fails to take into account firms that were unable to secure term loans and were granted short term lines of credit only. Wynant and Hatch found that 84 per cent of all businesses in the sample loan file had 'operating loans' (or lines of credit), as opposed to 48 per cent with term loans.[89] Operating loans 'provide the customer with a maximum amount of financing (called the authorized borrowing limit) for a period, usually a year, and during this period loans will fluctuate depending upon the firm's working capital needs'.[90] Some firms will have *both* term loans and lines of credit, while many others will have lines of credit only. The predominance of operating loans in the Wynant and Hatch loan sample, however, does indicate a tendency on the part of the banks to lend short.

Short term lending is a rational response to the high risks of lending to small businesses. Thus, it is not clear that the inability of some businesses to secure long term capital constitutes a 'capital gap' or infirmity in the market. Moreover, the long term debtload of small Canadian firms is higher than the long term debtload of foreign firms.[91] This suggests that Canadian banks do not have an inordinate bias towards 'short termism'.

Taking an equity participation

Taking an equity participation reduces the incentive of the equity holders to shift risk from debt to equity. Should redistribution occur, the equity participation allows the debtholder to recapture part of the amount appropriated. Canadian banks do not, however, use this technique to reduce the risks of opportunism.[92]

Mezzanine lenders, venture capitalists, and merchant bankers make extensive use of equity as a risk-reduction technique. Such lenders frequently service the riskier segment of the loan market – including high tech firms. They will typically lend at a high interest rate (on the order of 20–25 per cent) and take an 'equity kicker' with their debt. Examples of debt with an equity 'kicker' include a bond with warrant attached, and convertible debt.[93] Such lenders also rely less than banks on collateral and personal guarantees to secure their investments, and much more on monitoring and active participation in management.

Adjusting the loan yield

Loans to small businesses in Canada carry a higher interest rate than loans to larger (and generally less risky) businesses.[94] However, the interest rate curve (showing how interest rates vary as firm size increases) appears to be remarkably flat, and flatter than that in the United States.[95] While some of the flatness of the interest curve is accounted for by the fact that the banks have moved in the past decade to increasingly fee-based loan charges, taking this into account still leaves an unexpectedly flat interest curve. The flatness of the interest curve appears to be a product both of bank specialization in low-risk lending, and 'implicit regulation', under which the banks undercharge their higher risk customers and overcharge their lower risk customers. These topics are explored further below.

Refusing to lend

In the extreme, the lender might simply decide that the costs and risks of lending to small high tech businesses outweigh the expected gains. The higher risks and costs of lending to small, high tech firms would lead to the prediction that such firms are refused bank credit more often than other firms.

The 'accepted wisdom' suggests that this in indeed the case. A study which suggests otherwise, however, is discussed below.

Evidence concerning the riskiness of loans to small firms

Studies show that, in general, small firms tend to have higher beta coefficients – i.e. greater systematic risk – than larger firms.[96] Small businesses also have much higher earnings variability than do larger firms,[97] and higher failure rates.[98] These last factors reflect higher systematic and unsystematic risk.

In a recent study based on Canadian Federation of Independent Business (CFIB) survey results, Haines and Riding found that default rates on bank loans were highest (at 8.2 per cent) for firms with less than $500,000 in sales.[99] Curiously, however, they found that for firms with $501,000 to $1m in sales, the default was only 4.8 per cent – lower than for firms with from $1m to $2m in sales (7.6 per cent), and for firms with $2m to $5m in sales (6.6 per cent).

Evidence concerning the riskiness of loans to small, high technology firms

The Haines and Riding study used CFIB survey responses to test two hypotheses: (1) high tech firms experience greater difficulty than other small firms in securing bank financing (hypothesis 1); (2) high tech concerns pay higher interest rates than other small businesses, reflecting higher levels of risk (hypothesis 2).

The study found that neither hypothesis could be confirmed. In relation to hypothesis 1, turndown rates on loans to firms with 'medium tech' and 'high tech' products were no greater than those to firms with 'low tech' products.[100] In relation to hypothesis 2, there appeared to be little difference in the interest rate charged to firms classifying themselves as 'high tech' or 'medium tech', and those classifying themselves as 'low tech'.

There is some difficulty, however, in using the CFIB survey results to conclude that high technology concerns face no greater obstacles in securing bank financing than other businesses. The question put to respondents in the questionnaire was:

> Please evaluate the extent to which your firm's final product/service is 'low tech', 'medium tech' or 'high tech'. (A rating of 'low tech' means that your final product has little or no technology component, 'medium tech' means that there is a moderate amount and 'high tech' means the final product is exclusively 'high tech'.)[101]

The survey also asked respondents to:

> Please indicate below the extent to which your firm *employs high technology* to produce its final products or to provide its services. (A rating of 'low tech' means that your firm uses little or no high technology at all in production, 'medium tech' means that it uses a moderate amount of high technology and 'high tech' means that you use only high technology in production.)

The difficulties may be stated as follows:

1 The definition of 'high tech' is rudimentary in the extreme. Even sophisticated economists have trouble agreeing on what is meant by 'high technology', and self-classification on the basis of a very sketchy definition is likely to lead to inconsistent classifications.

2 The definition of 'high tech' in the CFIB survey does not capture the essence of innovative business. A firm might easily sell products with a 'technology component' but be purchasing those products from an oversees wholesaler. For example, a retailer importing VCRs from Japan and selling them in Canada might self-classify as selling a 'high tech' product. But the retailer is neither developing innovative technology, nor employing existing technologies in novel ways. In my view, only the latter type of firm deserves to be classified as 'high tech.' It is

firms that are doing basic R&D work, and *developing* new technologies, that are the backbone of the 'new' economy, and not firms that are merely selling products with a technology component – particularly if imported from elsewhere.

3 There may be a selection bias in the data. Respondents to the CFIB data were asked whether an application for credit was accepted or rejected by a financial institution.[102] This question may fail to capture informal turndowns where the bank manager indicated that there would be little likelihood of an application succeeding. Moreover, the turndown ratios fail to capture instances where applications were never made, because the prospects for success were perceived to be poor.

4 The sample of firms used consists exclusively of CFIB members, and will thus exclude start-ups and very small firms.

That the CFIB definition fails to capture the essence of what it is to be a high tech business appears to be confirmed by the industrial classifications selected by those engaging in 'high tech' business.[103] Of the 142 firms that identified themselves as 'high tech', 3 were engaged in 'Agriculture, Fishing, Farming', 19 were engaged in 'Mining, Primary Industries', 15 in 'Finance, Insurance, Real Estate', 4 in 'Business Services', 28 in 'Community Services', and 32 in 'Hospitality etc.' Some of these firms may be engaged in the development of novel technologies. For example, a biotechnology firm using genetic engineering to develop new crop strains would likely classify itself as involved in agriculture. However, one suspects that the majority of firms in the above-mentioned categories are probably not involved in the development of innovative technologies. It is more likely that they are employing 'high technology' devices, like computers or advanced machinery to produce their products. Indeed, on the basis of the CFIB survey, this chapter would likely be classified as a 'high tech' product, since it is being produced on a notebook computer. If this is indeed what is being captured by the survey results, then it does not appear to be a very useful definition of 'high tech'.

Firms that are doing R&D work (especially in the biotechnology/medical or other primary science fields) may spend years developing a product. In the interim such firms will have no sales, no accounts receivable, few collateralizable assets, and no positive cash flow. As indicated in *Financing the New Economy*,[104] even those banks which have been the most innovative in lending to high tech concerns require a finished product, a solid sales record, and positive cash flow to service debt interest. Thus, the CFIB survey results appear to fall short of supplying good evidence that high tech firms have no more difficulty in securing bank capital than other firms.

Evidence on the profitability of SME loans

While there appears to be no hard data on the profitability of SME loans,[105] Wynant and Hatch found that

> a significant portion of the account managers that we interviewed regard small business accounts as marginally profitable for their bank. Many perceive that small businesses borrow relatively small amounts of funds, are charged thin spreads above the bank's cost of funds, require considerable effort to monitor and manage, and account for a large share of the credit losses experienced on their bank's business loans.[106]

The study by Haines and Riding[107] offers supporting evidence for the view that profit margins on small loans are thin. Echoing findings by Wynant and Hatch, Haines and Riding find that in a random sampling of commercial loan files of the major chartered banks, 50 per cent of the loans were for amounts of $50,000 or less. They observe that if the loan rate is 3 per cent above prime, then the 'maximum annual loan revenue' is only about $2,000 per year:

> This result carries clear implications over to the bank–SME relationship. Assuming, for example, that on a loan of this small size banks assess an interest rate of prime +3, the maximum annual loan revenue from 50 percent of the bank's SME clients is approximately $2,000. After allowance for loan losses (typically estimated at 1 percent of balance), there remains little contribution to margin and overhead after the direct costs of the loan account managers are recognized. These economics drive the result that account managers must typically oversee 80 to 120 accounts. This result, in turn, implies contact time that averages 1–2 days per year.[108]

While neither of the above-mentioned studies focused on high tech businesses, there is every reason to believe (as sketched out above) that loans to high tech businesses will have even smaller profit margins than those to non-high tech businesses.

Sum: the comparative agency costs of debt and equity for small, high tech firms and the optimal debt/equity ratio

Pettit and Singer have suggested that as firm size shrinks, the agency costs of equity will rise faster than the agency costs of debt.[109] This is so, according to Pettit and Singer, because information asymmetry prevents insiders from adequately communicating the worth of the firm's investment opportunities to outside equity holders. Since these equity holders have a greater interest in the residual claim than debtholders, the cost of the information asymmetry is more serious for equity holders. This suggests that as firms get smaller, they are more likely to raise external financing in the form of debt capital.

This appears to paint an incomplete picture of the financing opportunities available to small, high tech firms. Such firms are characterized by considerable information asymmetry and little bankable collateral. Thus, opportunities to borrow money are scarce or non-existent.

In this segment of the market, capital is typically furnished by informal and formal venture capitalists, as well as mezzanine financiers and merchant bankers. These financiers are specialized high risk providers of capital. By virtue of their particular expertise, they are better able to resolve information asymmetries than are bankers. Moreover, they are willing to furnish capital in the absence of bankable collateral. Financiers of this character are often hybrid investors who take a combination of debt and equity capital, although in many cases they are quite willing to confine their participation to equity capital alone.[110]

It should also be recognized that monitoring capability and the ability to furnish collateral interact in determining whether the firm will seek (and obtain) debt or equity financing. Banks will tolerate a certain amount of information asymmetry so long as they can adequately secure their loans. It may well be the case, as Pettit and Singer suggest, that debt financing is cheaper than outside equity for many relatively low risk (non-high tech) small firms with available collateral.

LOAN RATES AND TERMS OF CREDIT FOR SMALL BUSINESSES: THE 'IMPLICIT REGULATION' HYPOTHESIS

Numerous studies have found that small businesses are charged higher rates of interest than larger firms.[111] There is also evidence, however, that the variation in interest charges between the smallest business accounts and the largest is small. A survey done for the Canadian Federation of Independent Business of CFIB members found that the average rate of interest charged by the banks was only 1.23 per cent above prime.[112] Similarly, in a representative sampling of Canadian bank loan files, Wynant and Hatch found that the average rate was 1.73 per cent above prime.[113] Moreover, nearly 80 per cent of both operating and term loans were charged less than prime plus 2 per cent.[114] A remarkable 95 per cent of operating loans, and 97 per cent of term loans were charged less than 3 per cent above prime.[115]

The interest spread alone, however, does not tell the whole tale. In the past ten years the banks have increasingly moved in the direction of charging their business clients loan application and loan management fees in addition to interest charges. Wynant and Hatch estimate that these fees add 88 basis points to the average premium above prime.[116] Thus, the Wynant study, corrected for loan fees, shows that the average premium above prime is approximately 2.6 per cent.[117] Moreover, the Wynant and Hatch study indicates that these fees 'can add significantly to the overall profit that a bank earns from a borrowing relationship'.[118]

While there is quite substantial variation in loan fees, however, 76.1 per cent of the sample were charged loan fees that added less than 1 per cent per annum to the total loan cost. Only 11 per cent of customers were charged loan fees that added more than 2 per cent to annual loan cost, and only 6.3 per cent were charged loan fees in excess of 3 per cent.[119]

It is virtually impossible to determine, a priori, whether SMEs are charged an economically 'fair' rate of interest for the degree of risk they present to the banks. However, the heightened risks and costs of servicing SME accounts, explored in detail below, give some reason to believe that riskier SME accounts are undercharged by the banks. The undercharging hypothesis is given support by a number of studies that have claimed that, because of sensitivity to criticism from government, the press, and small business organizations, the banks have unwritten rules under which they will rarely charge business customers in excess of 3.5 per cent above prime. For example, the Economic Council of Canada stated in 1982 that

> [t]his 'implicit regulation' means that the most 'visible' institutions assign a high cost to government sanctions and public criticism resulting from high prices. As a result, they prefer to resort to quantity rationing as opposed to price rationing. For instance, chartered banks will hardly ever charge an interest rate higher than prime plus 3.5 per cent. Because of the existence of a self-imposed ceiling on interest rates, funds will first go to customers presenting the lowest risk and to those whose accounts are least costly to manage. Small and medium-sized firms, as well as firms whose head offices are some distance from the major financial centres, do not meet these criteria and, consequently, are served last.[120]

While the Council appears to have been wrong in asserting that customers away from major financial centres are served last,[121] others have also claimed that the banks have self-imposed ceilings on the interest rate that they will charge.[122]

While credit rationing by itself does not mean that loan customers who are accepted for loans will be undercharged, it is not unreasonable to suppose that the same political pressures that lead the banks to engage in credit rationing might also lead them to undercharge small business customers. Indeed, coming out of the most recent recession, many complaints were heard from small businesses, organizations representing small business, and politicians about the drying up of bank credit for small business. In fact, a study by Haines and Riding casts considerable doubt on the view that credit has become less available to some businesses.[123] None the less, the government has responded to political pressure by in turn putting pressure on the banks to make loans to small business customers. In response, some of the major banks have announced that they will double their Small Business Loans Act (SBLA) activity, in addition to lending to small business customers covered by the SBLA at or close to prime, and indeed in some cases at less than prime.[124] It is difficult to imagine that charging these risky customers the prime rate could be a prudent or efficient pricing strategy.

Orser *et al.* produce direct evidence that banks engage in credit, rather than price rationing. 'Micro' businesses (with fewer that three full time employees and annual sales of less than $200,000) as well as 'mini' businesses (with 3–5 full time employees and between $200,000–500,000 annual sales) were found to experience greater difficulty in getting access to credit than larger firms.[125] However, when such firms were successful in securing credit, the interest rate charged was found to be close to prime, and the rate did not vary substantially between micro and mini businesses.[126] The study also found evidence that default rates for micro and mini businesses were no higher than for larger small businesses.[127] Orser *et al.* concluded that their evidence was consistent with the view that the banks ration credit to less risky small business ventures, rather than simply charging riskier ventures a higher interest rate. Their evidence is consistent with the view that banks undercharge their smallest business borrowers.

Haines *et al.* provide further evidence in support of both credit rationing and implicit regulation.[128] The Haines study found that when a firm had experienced financial distress it was more likely to be turned down for a term loan. However, firms that had experienced financial distress and were granted term loans were charged no higher rates of interest than were firms that had not experienced financial distress. This finding is consistent with credit, but not price rationing.[129]

The implicit regulation hypothesis is given a further boost by a recent study that compared loan pricing by Canadian and US lenders.[130] The study found that Canadian lenders charged almost the same interest rate to a wide range of borrowers by risk classification. US lenders charged higher risk customers about 70 basis points more than Canadian lenders, and charged lower risk customers about 20 basis points less.[131]

The study found a similar picture when customers were arrayed by size. While US and Canadian lenders charged their largest customers (in excess of $250 million in sales) about the same interest rate, US lenders charged lenders with less than $50 million in sales a full percentage point more than Canadian lenders.[132]

While the study does not look at 'small' and 'medium' sized firms as usually defined, the results are none the less both interesting and puzzling. It may be that Canadian banks charge much higher loan application and loan management fees than their US counterparts: the study did not compare differences in fees. The results may also suggest, however, that implicit regulation operates in the realm of large firms as well as small.

There is an alternative, efficiency-based explanation for why banks might lend only to less risky business borrowers. Because of the difficulty in training loan officers to evaluate a wide range of risks, it has been suggested that it is optimal for lenders to limit those segments of the market that can be served by any particular loan officer.[133] Stiglitz and Weiss theorize that in the face of poor information about credit risks, it may be optimal for lenders to refuse to lend at high interest rates even to customers who are willing to pay high rates.[134] Stiglitz and Weiss posit that there exists an optimal lending rate. Loaning money to willing customers at higher rates actually diminishes bank profits. Willingness to lend at high interest rates attracts bad credit risks, creating an adverse selection effect that is worse than that experienced on loans at the lower rate.[135]

Moreover, charging a higher interest rate may perversely affect the behaviour of borrowers, causing them to invest the borrowed funds in high risk and negative net present value projects.

Under this view (which might be styled the 'task specialization' view), lenders should specialize in markets in which they have a comparative advantage in assessing credit risk. Thus, we should observe different institutions either servicing different risk segments of the loan market, or establishing a variety of independent lending facilities to service different segments of the loan market.

There is evidence that the major banks have decided to pursue the latter course. Many of the banks now have merchant banking subsidiaries that make subordinated loans at much higher interest rates than the banks themselves. The loan officers at these facilities are significantly more educated than loan officers at the banks.[136] These lending facilities are removed from the banks' main business by placing them in separate corporate subsidiaries. This allows the banks to be players in the high risk debt market while strictly limiting their risk exposure. It also allows them to exploit the benefits of loan manager task specialization in more than one market.

Of course, there are many independent players specializing in the high risk market, including merchant banks, venture capitalists, and mezzanine lenders. Thus, there seems to be support for an efficiency-based explanation of bank credit rationing and interest rate ceilings.

In my view, *both* the implicit regulation and task specialization hypotheses partly explain bank lending policy *vis-à-vis* small business.

BANKS AND HIGH TECHNOLOGY COMPANIES: NEW HORIZONS

Industry Canada's recently published study, *Financing the New Economy*, under the leadership of banker Paul Toriel, indicates that it is possible for banks to play a role in financing high technology concerns. NatWest Bank in England and The Silicon Valley Bank in California have both adopted the innovative approach of partnering with a venture capital firm. The venture capitalist takes primary responsibility for the 'due diligence' on a prospective client (i.e. investigation into the soundness of the business plan, the prospect for growth and profitability, creditworthiness, etc.). Once the venture capitalist has established a successful financial relationship with the firm, it will refer the firm to the bank to receive a part of its financing in the form of a loan.

The NatWest/Silicon Valley model has inspired the major Canadian banks to make attempts to borrow from this model. However, while it shows great promise, it should not be forgotten that it works only with firms with a proven product, an established record of sales, and significant positive cash flow. The model is not, nor will it likely ever be, applicable to early stage firms that are engaged solely in product development and lack both a product and positive cash flow. Moreover, to date, no Canadian bank has adopted wholesale the NatWest/Silicon Valley model.

POLITICAL DIMENSIONS

Do the banks have a duty to fund SMEs?

The Berger Committee was not content with the view that banks should simply maximize profits. According to the Committee:

> Financial institutions, especially the banks, in accepting that their own responsibility to society is greater than 'merely' being efficient, stable and profitable, must recognize the importance of small and

> medium-sized businesses to the economy . . . The Committee believes that the banks have a responsibility to serve small business which stems from the privileged position they occupy in the Canadian financial services industry. The branch banking system gives them access to a huge deposit base, a relatively cheap source of financing. They are the dominant lenders to small and medium-sized businesses, accounting for about 90% of lending to the small business sector.[137]

The Committee criticized the banks for failing to meet this responsibility, stating that:

> While the Committee understands that we have been in a recession, during which the call on credit would naturally go down, we believe that the actions of the banks have exacerbated the situation. Evidence gathered informally, as well as in the letters and testimony received, strongly suggests that banks called in lines of credit in an attempt to get their houses in order. The banks allege that the evidence of small and medium-sized business borrowers is 'anecdotal', by which they suggest it may be unreliable or inaccurate because confidentiality restricts their ability to respond. We do not agree. It is not credible that so many small and medium-sized business borrowers would imagine a false situation.[138]

This conclusion was arrived at despite the fact that the Committee had available to it a thorough study by Haines and Riding offering strong support for the view that there was in fact *no* 'credit crunch', and that reductions in bank lending to SMEs in the 1990–4 period were caused by demand and not supply-related factors.

While laudable in many other respects, it is difficult to understand why the Berger Report chose to prefer anecdotal evidence offered by self-interested small business people over the much more systematic and dispassionate evidence of Haines and Riding. Indeed, Haines and Riding point out that in survey data asking small business respondents if they are satisfied with their banking relationships there is a 'non-response' bias. Those who are content are far less likely to respond than those with complaints, biasing the data. Precisely the same reasoning is applicable to appearances before Parliamentary Committees. Those who are content with their banking relationships are likely to feel little compulsion to testify. Thus, those who appear before such a Committee are likely to be an unrepresentative sampling of bank borrowers. Both those with legitimate grievances, and pure cranks, are almost certain to be overrepresented.

Extrapolating even from legitimate grievances, when presented anecdotally rather than systematically, is a dangerous practice. Regardless of how they are frequently portrayed in the press, the banks are not faceless, soulless money machines supervised by a giant computer. They are staffed by people. It should not go without comment that people sometimes make mistakes. Indeed, in institutions as large as the major banks, mistakes in making or calling loans are unavoidable. It would thus be astonishing if, at any given moment in time, a subset of the banking population did not have legitimate grievances against the banks.

But this does not mean that *the system* is malfunctioning. In the end, anecdotal evidence is of far less probative value than systematic studies of the nature of the loan market, particularly when (as properly brought to the Committee's attention) the banks are unable to respond with countervailing anecdotal evidence due to confidentiality concerns.

Moreover, there is no reason to believe that banks have any special responsibility to finance SMEs. Banks are private, not public enterprises. Their duty is to maximize profits for shareholders, within the framework of commonly accepted business ethics and a regulatory structure that seeks to protect depositors' capital. If the government believes that there truly are 'capital gaps', and that some segments of the market are underfunded, then it is the responsibility of the government, not private citizens (whether individual or corporate), to correct the situation. The government has a variety of instruments at its disposal to effect this end, such as through loan subsidization or guarantee programs like the SBLA, lending/venture capital institutions like the Business Development Corporation (formerly the Federal Business Development Bank), tax incentives to funds investing in small businesses (or tax credits to those who invest in such funds), such as the Labour Sponsored Venture Capital Corporations, or by other means. But it is unfair and unreasonable to put this burden on the shoulders of for-profit lenders.

The tenor of the Berger Report offers further support to the 'implicit regulation' view described above. In addition to criticizing the banks on the basis of questionable anecdotal evidence, the Committee has indicated that it will call the banks to account on a periodic basis in relation to small business financing. This sort of strong-arm government tactic is, in my view, inappropriate and unwarranted.

Conclusion

While this chapter is titled 'The Banks and Innovative Enterprise: Opportunities and Constraints', the focus has clearly been much more on constraints than opportunities. Paul Toriel's chapter is an extraordinarily helpful glimpse at some of the opportunities. None the less, in my view, while banks have a role to play in financing innovative enterprise, the opportunities should not be overstated. Banks are low risk lenders who have traditionally lent against collateral or receivables. In the early stages, while research is underway and products merely under development, many high tech firms lack either hard assets or receivables to serve as collateral. Moreover, most bank loan officers are simply not in a position to evaluate high tech businesses. While the banks have proved themselves capable of lending to innovative firms with a proven product and a solid sales record, it is doubtful that they will ever be in a position to fund early stage high technology business.

NOTES

1 The term 'bankable' means that a firm satisfies bank lending criteria.
2 Banks are not the only institutions that make loans to small businesses in Canada. Other institutions, such as credit unions, caisse populaires, and trust companies do so as well. However, the 'big 6' chartered banks dominate commercial lending in Canada, making approximately 80 per cent of all private commercial loans. See Jerry Zeidenberg, 'Shaking the Money Tree' (June 1993), *The Globe and Mail Report on Business Magazine* 69, and see also George Haines, Allan Riding and Roland Thomas, 'Small Business Turndowns by Canadian Chartered Banks: Some Empirical Findings', Carleton University, School of Business Working Paper Series, WPS 89-06 (1989). Further, other private lending institutions would appear to use lending criteria similar to those used by the banks. Thus, throughout, I will refer only to banks.
3 See Larry Wynant and James Hatch, *Banks and Small Business Borrowers*, The Western Business School, The University of Western Ontario, London, Canada, 1991, at 5, 160–1; Haines *et al.*, *supra*, note 2; Lola Fabowale, Allan Riding and Catherine Swift, 'Gender, Structural Factors, and Credit Terms Between Canadian Small Businesses and Financial Institutions', Carleton University School of Business, WPS 92-1.
4 Wynant and Hatch, *supra*, note 3, at 164.
5 Ibid.
6 See EEC Study, *Intervention and Efficiency*, Economic Council of Canada, Ottawa, Canada, 1982, at 25 (smaller businesses generally have fewer collateralizable assets); Patricia Johnston, *Nothing Ventured . . . Investing in Canada's Winners*, The Toronto Stock Exchange, 1980, at 18 ('Smaller firms have roughly half the fixed assets of their larger rivals . . . The more knowledge-intensive firms have even lower ratios . . .' See also p. 34 (study of ten small firms showing a ratio of fixed to total assets of 20 per cent, versus a 30 per cent average for SMEs generally and 60 per cent for larger firms); Wynant and Hatch, *supra*, note 3, at 193 (soft assets like computer software are not acceptable as collateral).
7 Premier's Council, *Competing in the New Global Economy*, Report of the Premier's Council, Province of Ontario, 1988, vol. 2, at 256. While a long lag between initial research and development and product marketing tends to characterize high tech enterprise, this is not universally true. Software development cycles, for example, can be quite short.
8 Wynant and Hatch, *supra*, note 3, at 161.

9 Wynant and Hatch, ibid., at 19, 163. The Wynant and Hatch study indicates that the large Canadian banks took security over software in only 0.1 per cent of all loan files examined. Ibid. at 164. See also 'Software Companies Battle With Banks', *The Globe and Mail*, 27 December 1989, B1.

10 Wynant and Hatch, *supra*, note 3, at 9. The authors note that:

> Although bankers usually take collateral security with a value seemingly well in excess of the firm's loan facility, the bank seldom recovers its total funds in those cases where the security is realized upon. The assets pledged by small businesses as collateral usually command a value in a liquidation that is only a fraction of their recorded book value or the value estimated by the principal.

Ibid,. at 9. See also 165–7. See also Michael Trebilcock, Marsha Chandler, Morley Gunderson, Paul Halpern, and Jack Quinn, *The Political Economy of Business Bailouts* (Ontario Economic Council, Toronto, 1985), at 126; Facsym Research Limited, *Small Business Financing and Non-Bank Financial Institutions* (1982).

11 One of the reasons why recoveries are so low is that 'assets are sometimes drained off by the principals as the business encounters increasingly more difficult problems and eventually fails.' Wynant and Hatch, *supra*, note 3, at 167. Assets of high tech firms (e.g. software, formulas, etc.) are likely to be more easily moved than assets of larger firms. Realizations are also low because specialized equipment or inventory has low resale value. Ibid. Because of transaction-specificity, assets of high tech firms are likely to be particularly subject to this problem.

12 The value of small firms assets tends to be dissipated very quickly as the firm approaches and ultimately enters insolvency. See Trebilcock *et al.*, *supra*, note 10; Wynant and Hatch, *supra*, note 3, at 166.

13 See generally Oliver Williamson, 'Corporate Finance and Corporate Governance' (1988) 43 *J. Fin.* 567.

14 It may be that some early stage high tech firms have non-bank debt financing, such as vendor financing, and are relatively highly levered for this reason. Firms financed in this manner will tend to have high systematic risk and thus be unattractive borrowers for banks.

15 Stephen A. Ross and Randolph W. Westerfield, *Corporate Finance* (St Louis: Times Mirror/Mosby College Publishing, 1988), at 255–6.

16 See Premier's Council, *supra*, note 7, vol. 1, at 170, and vol. 2, 256–7.

17 Premier's Council, *supra*, note 7, vol. 1, at 139, 169; vol. 2, at 256–7. Product development times will vary greatly as between different sectors of the high tech market. Software can be developed in a period of a few months. However, the average time from synthesis of a drug molecule to marketing the drug is in the range of fifteen years. See Henry I. Miller, 'Desperately Seeking an FDA Commissioner', *Chicago Tribune*, 9 January 1997, p. 25.

18 Wynant and Hatch, *supra*, note 3, at 39.

19 Ibid., at 125. Historical loss rates have been close to these targets. Ibid., at 126. While President of the Canadian Bankers' Association, Helen Sinclair also indicated that the target loan loss rate of the banks was 0.5 per cent. See Helen Sinclair, Speech to the North Saskatoon Business Association (available from the Canadian Bankers' Association).

20 Sinclair, *supra*, note 19; Wynant and Hatch, *supra*, note 3, at 3, 12–13, 23, 124–7; D. Thornton, 'Small Business Financing Policies and Practices: An

Interview Survey of Managers', in *Small Business and Non-Bank Financial Institutions* (Toronto: Facsym Research, 1981).

21 In general, loan officers are instructed to evaluate prospective loan customers on the basis of the 'Five Cs' of commercial lending: Capital, Collateral, (Business) Conditions, Capacity, and Character. See, e.g. Lola Fabowale, Allan Riding, and Catherine Swift, 'Towards a Multivariate Model of Factors Influencing Terms of Bank Credit', presented at 1991 World Conference, ICSB, Vienna.

22 Wynant and Hatch, *supra*, note 3, at 125.

Although some of the banks have recently been more aggressive in pursuing high technology loan business, loans are none the less made only to high technology businesses with a mature product and an established market, reflecting this inherent conservatism. See Paul Toriel, *Financing the New Economy: Towards a Positive Conspiracy*, Project Report, Industry Canada, June 1994; Rick Lunny, 'Conventional Bank Financing for an Unconventional Industry: Lending to the Leading Edge', Investing in Information Technology Companies Conference, Toronto, 1 March 1994.

23 See, e.g. A. Riding, P. Dal Cin, L. Duxbury, G. Haines, R. Safrata, *Informal Investors in Canada: The Identification of Salient Characteristics*, A Report submitted to the Federal Department of Industry, Science and Technology Canada and to the Ministry of Economic Development and Trade of the Province of Ontario, 23 May 1993, at 27 (most common reason for rejection of investment opportunity by informal venture capital investors is lack of faith in the quality of management).

24 The non-human capital might be sold back to the entrepreneurs (who likely will value it more highly than anyone else). However, the same wealth constraint that led the entrepreneurs to seek outside funding in the first place may limit the entrepreneur's ability to repurchase these assets, particularly if the firm has failed and the entrepreneur's personal capital has been depleted.

25 See, e.g. ECC Study, *supra*, note 6, at 26.

26 R. Richardson Pettit and Ronald F. Singer, 'Small Business Finance: A Research Agenda' (1985) 14 *Fin. Management* 47, at 55.

27 Ibid., at 52.

28 Ibid.; Trebilcock *et al.*, *supra*, note 10, at 127.

29 Trebilcock *et al.*, *supra*, note 10, at 127.

30 The enumerated factors also affect equity investors.

31 See, e.g. *Dodge* v. *Ford Motor Company*, 204 Mich. 459, 170 N.W. 668, 3 A.L.R. 413 (1919).

32 See, e.g. Amir Barnea, Robert A. Haugen and Lemma W. Senbet, *Agency Problems and Financial Contracting* (Prentice-Hall, Englewood Cliffs, New Jersey, 1985).

33 P. J. Halpern, M. Trebilcock and S. Turnbull, 'An Economic Analysis of Limited Liability in Corporation Law' (1980), 30 *U. Tor. L. J.* 117; A. Barnea, R. Haugen, and L. Senbet, 'Market Imperfections, Agency Problems, and Capital Structure: A Review' (Summer 1981) *Fin. Management* 7. Contrary to Jensen and Meckling, I assume that the amount of debt remains constant while the proportion of inside equity ownership rises. See Michael Jensen and William Meckling, 'Theory of the Firm: Managerial Behavior, Agency Costs and Ownership Structure' (1976), 3 *J. Fin. Econ.* 305 (assuming that as inside equity increases, debt ownership proportionately decreases, reducing the incentive of the equity holders to appropriate wealth from debtholders and hence reducing the agency costs of equity).

34 Trebilcock *et al.*, *supra*, note 10. This is consistent with the view that smaller firms can more easily adjust their asset portfolios to meet new markets.

35 See, e.g. John C. Coffee, 'Shareholders Versus Managers: The Strain in the Corporate Web' (1986), 85 *Mich. L. Rev.* 1.

36 Ibid.

37 There is some reason for caution in arriving at this conclusion. Even when the firm fails, the manager may have learned valuable lessons from the failure which may be used the next time s/he embarks on a similar enterprise. In other words, the very experience of failure may impart valuable human capital to the entrepreneur, offsetting the loss of human capital in the failed enterprise.

38 As indicated above, small high tech firms are typically characterized by high levels of unsystematic risk. It may be that at high levels of risk, increases in risk will have a greater marginal cost for managers. If so, the managers of such firms will be more reluctant to increase risk than the managers of larger (and less risky) firms.

39 I assume here that the key shareholders and the managers are the same people, which is usually the case in a small firm.

40 Stripping vital assets out of the business will obviously imperil its existence. Thus, so long as the entrepreneur remains committed to the success of the enterprise, asset stripping is unlikely to occur. It is when the enterprise approaches insolvency that the most serious danger of asset stripping arises.

41 Halpern *et al.*, *supra*, note 33; Pettit and Singer, *supra*, note 26.

42 Wynant and Hatch found that personal guarantees were required in 87 per cent of small business loans, although pledges of personal assets were required only one-third of the time. See Wynant and Hatch, *supra*, note 3, at 8–9, 165–7.

43 Of course, a personal guarantee will also address other forms of redistribution from creditors to shareholders.

44 Or by the entrepreneur/managers, who will consider how the project affects their welfare both as equity holders and managers. Even if the project requires no fresh infusion of capital, the managers must expend effort to develop a new project. That effort may result in no additional compensation or benefit to equity holders.

45 Trebilcock, *et al.*, *supra*, note 10, at 139.

46 Where subordinate debt claims are issued, however, the capital raised may operate as a financial cushion against receivership or insolvency, enhancing the value of superior debt claims. Whether the raising of additional capital pushes the firm closer to, or farther away from insolvency will depend on the firm's cash flow and its ability to service the new debt.

47 Thornton, *supra*, note 20; Fabowale *et al.*, *supra*, note 21 ('depth of financial management and the track record of the firm seem to be the variables which are most closely related with loan turndowns').

48 See, e.g. Wynant and Hatch, *supra*, note 3, at 15, 26 (finding that many small firm managers did not even understand the terms or cost of the loan facility).
Wynant and Hatch consider the absence of financial management skills in small business owners sufficiently serious to warrant a recommendation that the government make funds available for imparting management skills to small business managers. Ibid., at 26–8. The Business Development Corporation (formerly the Federal Business Development Bank) now provides such services.

49 Barnea *et al.*, *supra*, note 33.

50 See, e.g. CBCA s. 163.
51 Wynant and Hatch, *supra*, note 3, at 14, 17.
52 Premier's Council, *supra*, note 7, vol. 1, at 172. In a similar vein, Mary Macdonald states that

> the ability of the management team will be of overriding concern throughout the whole process [of selecting an investment]. The biggest mistake an entrepreneur can make in the search for venture capital is assuming the investor is backing the technology or product, rather than the management team.

Mary Macdonald & Associates, *Venture Capital in Canada: A Guide and Sources*, Association of Canadian Venture Capital Companies (1992 edition), at 23. See also Dominique M. Short and Allan M. Riding, 'Informal Investors in the Ottawa–Carleton Region: Experiences and Expectations' (1989), 1 *Entrepreneurship and Regional Development* 99, at 108. Informal interviews with venture capitalists (and others involved in small firm financing) yielded a similar view.

53 The section immediately preceding also emphasized that poor management is a risk faced by lenders. This section stresses that outsiders such as lenders face difficulty in evaluating managerial skills in order to determine whether such managers are indeed poor managers.

Management risk is not unique to private financing in the early stages of development. One factor that apparently keeps many small firms from going public is the perception that they lack good quality management. See Johnston, *supra*, note 6, at 35. In a similar vein, a Vice-President and Director of RBC Dominion Securities Inc. recently wrote that 'our experience reveals that the most common reason for IPO's [initial public offerings] not working out is the quality of management'. See Norman Fraser, 'Going Public: The Underwriter's Perspective', *Insight*, 22 June 1993, at 3.

54 Entrepreneurs are likely to be more risk preferring than the average, although whether the variance in risk preference is higher for entrepreneurs than for the population at large is not known.
55 See, e.g. Fabowale *et al.*, *supra*, note 3 ('depth of financial management and the track record of the firm seem to be the variables which are most closely related with loan turndowns'); Barbara J. Orser, Allan J. Riding and Catherine Swift, 'Banking Experiences of Canadian Micro-Businesses', unpublished, June 1993 (the presence of a financial professional in the firm makes a loan turndown less likely); Wynant and Hatch, *supra*, note 3, at 7 (small companies with managers with limited experience refused for loans more often). Indeed, 'character' is one of the '5 C's' of commercial lending. See, e.g. Fabowale *et al.*, *supra*, note 21.
56 Wynant and Hatch, *supra*, note 3, at 17.
57 Ibid., at 20.
58 M. J. Grant & Co. Ltd., *Small Business Views the Banks: The Bottom Line*, Canadian Federation of Independent Business, 1988.
59 See also W. Dunkelberg, J. Scott and E. Cox, 'Small business and the Value of Bank–Customer Relationships' (1984), 14 *J. Bank Research* 248, for evidence that bank managers do not know their business customers very well.
60 Wynant and Hatch, *supra*, note 3, at 119.
61 Ibid., at 119. In fact, Wynant and Hatch's 1991 survey found that only 40 per cent of account managers had university degrees. Ibid., at 120.

62 *Focus 2000: Report of the Task Force on Making Investment Capital Available*, The Canadian Chamber of Commerce, August 1988 (Lloyd R. McGinnis, Chairman) at 30; Premier's Council, *supra*, note 7, vol. 1, at 175–7; John Southerst, 'The Start-Up Star Who Bats .900' (March 1993), *Canadian Business* 66.
63 Wynant and Hatch, *supra*, note 3, at 17.
64 Ibid., at 20.
65 Ibid., at 160.
66 See, e.g. Michael C. Jensen, 'The Takeover Controversy: Analysis and Evidence', in *Knights, Raiders and Targets: The Impact of the Hostile Takeover*, John C. Coffee, Jr *et al.*, eds. (New York: Oxford University Press, 1988), at 314.
67 Randall Morck and David A. Strangeland, 'Corporate Performance and Large Shareholders', Working Paper No. 4-94, Institute for Financial Research, Faculty of Business, University of Alberta.
68 Similarly, Slovin and Sushka find that stock prices of publicly traded firms tend to rise on news of the death of a significant individual blockholder. M. Slovin and M. Sushka, 'Ownership Concentration, Corporate Control Activity, and Firm Value: Evidence from the Death of Inside Blockholders' (1993), 48 *J. Fin.* 1293.
69 Eugene F. Fama, 'Agency Problems and the Theory of the Firm' (1980), 88 *J. Pol. Econ.* 288.
70 In larger firms, the larger size of the managerial 'team' will make it more difficult to evaluate the performance of individuals within the team.
71 The Premier's Council reports that many entrepreneurs fail once or twice before ultimately succeeding. Premier's Council, *supra*, note 7, vol. 1, at 182. This might demonstrate a learning curve, however, as much as it demonstrates early bad luck.
72 Frank H. Easterbrook, 'Two Agency Cost Explanations of Dividends' (1984), 74 *Amer. Econ. Rev.* 650.
73 Michael C. Jensen, 'Agency Costs of Free Cash Flow, Corporate Finance, and Takeovers' (1986), 74 *Amer. Econ. Rev. (Papers & Proceedings)* 323.
74 Moreover, the entrepreneur/manager will typically hold a much higher percentage of the firm's equity and will thus have a much greater interest in seeing the firm succeed.
75 In the Premier's Council survey of 71 small Ontario companies, 74 per cent identified access to equity capital as a constraint facing the firm. Premier's Council, *supra*, note 7, vol. 1 at 169; vol. 2 at 259. This was a higher percentage than for any other constraint identified. Further, 'lack of equity capital' was the most commonly identified constraint for all three types of firms surveyed ('Struggling', 'Growing', and 'Exceptional' firms). Ibid., vol. 2, at 260. 'Lack of debt capital at an affordable cost' was the second most commonly identified constraint, with 59 per cent of small businesses stating that this was a problem. Ibid., vol. 1, at 169. More firms (54 per cent) identified 'lack of equity capital' as a 'major constraint' than any other type of constraint. 'Lack of Debt capital at an affordable cost' was the second most commonly identified 'major constraint', at 29 per cent. Ibid., vol. 2, at 259.
76 Wynant and Hatch, *supra*, note 3, at 164.
77 Ibid.
78 Ibid., at 160.
79 Ibid., at 165.
80 Ibid., at 8–9, 165–6.

81 See generally Smith and Warner, 'On Financial Contracting: An Analysis of Bond Covenants' (1979), 7 *J. Fin. Econ.* 117.
82 Wynant and Hatch, *supra*, note 3, at 173.
83 Ibid.
84 Pettit and Singer, *supra*, note 26, at 52, 55.
85 Wynant and Hatch, *supra*, note 3, at 173. The Wynant and Hatch loan sample is limited to loans of $500,000 or less. Their survey of typical loan covenants should be used with some caution, as below this threshold the survey does not stratify firms by size, but rather lumps together all loan sizes. This will obscure differences that arise within the sample.
86 See, e.g. Johnston, *supra*, note 6, at 18.
87 Wynant and Hatch, *supra*, note 3, at 162.
88 Ibid.
89 Ibid., at 123.
90 Ibid., at 123.
91 See Jeffrey G. MacIntosh, 'Legal and Financial Barriers to Financing Innovative Enterprise in Canada', Government and Competitiveness Project, School of Policy Studies, Queen's University, 1994, at 6.
92 Since this chapter was written, the Canadian banks have begun to explore the possibility of using equity features (such as warrants) in their loans to high-growth high-risk businesses.
93 See Barnea *et al.*, *supra*, note 33, at 93–7. 'Strip financing', in which the lender takes comparable amounts of debt and equity, serves a similar purpose.
94 See, e.g. Fabowale *et al.*, *supra*, note 3; Fabowale *et al.*, *supra*, note 21; B. Orser, A. Riding and C. Swift, 'The Face of Micro Business in Canada: Perceptions and Realities', presented at June 1993 ICSB Conference, Las Vegas.
95 See pp. 140–4.
96 Pettit and Singer, *supra*, note 26, at 53.
97 ECC Study, *supra*, note 6, Table 3–4, at 25.
98 For example, the Premier's Council found that the failure rate of companies with fewer than five employees was 40 per cent per year, versus 4 per cent for companies with over 500 employees. Premier's Council, *supra*, note 7, at 166, and vol. 2, at 249. The ECC study also reported that 'The highest proportion of firms with both a negative rate of return and a high debt-to-asset ratio is found among businesses with assets of less than $250,000.' ECC Study, *supra*, note 6, at 24. By contrast, Orser *et al.*, *supra*, note 94, found that firms employing fewer than five people (or one person full time), had no higher default rates on bank loans than larger small businesses. The later finding, however, is consistent with the theory that banks ration credit to the most creditworthy borrowers. See *infra*.
99 George Haines, Jr and Allan L. Riding, 'Recent Experience with the SBLA: Economic Impacts, Incrementality and Risk Profile Analysis', Submitted to Industry Canada, 17 August 1994, at 15.
100 George Haines, Jr and Allan L. Riding, 'Access to Credit: Lending Priorities and SMEs', Submitted to Industry Canada, 17 August 1994, table 2. Turndown rates were slightly higher in respect of applications for new lines of credit, but the difference was not great. See also Allan Riding and Catherine Swift, 'Banking on Technology: The Environment for Bank Borrowing by Small Canadian Technology-Based Firms' (1993) *J. Small Bus. and Entrepreneurship 13* (arriving at a similar conclusion).
101 Haines and Riding, *supra*, note 100, Appendix A. See also Riding and Swift, *supra*, note 100.

102 Haines and Riding, *supra*, note 100, Appendix B.
103 Ibid., Tables 9 and 10.
104 Toriel, *supra*, note 22.
105 Wynant and Hatch, *supra*, note 3, at 129.
106 Indeed, a recent industry publication indicates that all of the major banks have committed a large amount of capital to funding small businesses under the SBLA program, at rates below the SBLA ceiling. See *The Banking Industry Supports the Small Business Sector*, Canadian Bankers Association, 25 January 1994. It is difficult to imagine that such business is very profitable.
Curiously, Wynant and Hatch, *supra*, note 3 also found that 'senior management in the six banks view the small business sector as an important and profitable market for their institutions'. Ibid., at 128. It may be that, to some extent, senior managers are anxious to give the 'politically correct' view of small business lending.
107 Haines and Riding, *supra*, note 100.
108 Ibid., at 41.
109 Pettit and Singer, *supra*, note 26, at 56ff.
110 Jeffrey G. MacIntosh, 'Venture Capital Exits in Canada and the United States', in Paul J. N. Halpern, ed., *Financing Growth in Canada* (Calgary: University of Calgary Press, 1997), 279–356.
111 See, e.g. Fabowale *et al.*, *supra*, note 3; Fabowale *et al.*, *supra*, note 21; Orser *et al.*, *supra*, note 55; Wynant and Hatch, *supra*, note 3.
112 M.J. Grant & Co. Ltd., *supra*, note 58, at 34.
113 Wynant and Hatch, *supra*, note 3, at 158.
114 Ibid., at 157.
115 Ibid., at 158.
116 Ibid., at 157.
117 Wynant and Hatch, *supra*, note 3, note that the average premium above prime in the 1991 study exceeded the 1982 average by 80 basis points, which is almost exactly the effective increase in interest rates resulting from service fees. This demonstrates the shift from interest only to interest plus service fees.
118 Ibid., at 157.
119 Ibid., at 159.
120 ECC Study, *supra*, note 6, at 26. Both the concept and the term 'implicit regulation' appear to have first been used by the Economic Council of Canada. The term has also been used, however, by Jensen and Murphy in the context of executive compensation. See Michael C. Jensen and Kevin J. Murphy, 'Performance Pay and Top Management Incentives' (1990), 98 *J. Pol. Econ.* 225.
121 In fact, firms in urban areas appear to experience higher loan turndown rates. See, e.g. Fabowale *et al.*, *supra*, note 21; Haines *et al.*, *supra*, note 2; M. J. Grant & Co. Ltd, *supra*, note 58, at 32.
122 See, e.g. James Hatch, Larry Wynant, and Mary Jane Grant, *Government Loan Guarantee Programs for Small Business* (School of Business Administration, University of Western Ontario, London, Ontario, 1985), at 3 (the banks place 'ceilings on the credit risks that are acceptable and correspondingly, on the loan rates it will charge').
123 Haines and Riding, *supra*, note 100.
124 See *The Banking Industry Supports the Small Business Sector*, Canadian Bankers Association, 25 January 1994. Under the SBLA, the government guarantees up to 90 per cent of the loan. Banks are allowed to charge an interest rate up to prime plus 1.75 per cent.

125 Orser, *et al.*, *supra*, note 55; (loan turndown rates related to size of firm). See also Orser *et al.*, 'The Face of Micro Business in Canada: Perceptions and Realities', *supra*, note 94.
126 Orser, *et al.*, *supra*, note 55, at 13 (interest rates on the order of 1.5–2 per cent above prime; interest rates for mini businesses in most cases no more than half a percentage point higher than for micro businesses).
127 Further evidence may be found in Orser *et al.*, *supra*, note 94. This last finding stands in sharp contrast to the finding of the Premier's Council that the failure rate of micro-businesses generally is 40 per cent per year, versus 4 per cent for companies with over 500 employees. Premier's Council, *supra*, note 7, vol. 1, at 166; vol. 2, at 249.
128 Haines *et al.*, *supra*, note 2.
129 Ibid. These firms did, however, have interest rates 64 basis points higher, on average, for lines of credit.
130 Christopher L. Snyder, 'Commercial Loan Pricing in Canada . . . The First View is Worrisome', Loan Pricing Corporation (as reprinted in Burns Fry Limited client newsletter), 29 March 1993.
131 Ibid., at 1.
132 Ibid., at 2.
133 See, e.g. Facsym Research Limited, *supra*, note 10, at 15 ('[l]oan officer training and experience equips them to appraise a limited variety of risks').
134 J. E. Stiglitz and A. Weiss, 'Credit Rationing in Markets with Imperfect Information' (1981), 71 *Amer. Econ. Rev.* 393. See also Facsym Research Limited, *supra*, note 10, at 15–16; Trebilcock, *supra*, note 10, vol. 1, at 134.
135 'Adverse selection' was first noted in insurance markets. A company selling house insurance, for example, might initially set rates based on the average loss experience in the population as a whole. However, at this price, those experiencing a higher risk of loss will find the insurance a bargain, and will buy the insurance in numbers disproportionate to their representation in the population (while those with a low risk of loss will be disinclined to insure). The population of insureds will then be biased towards those with a high risk of loss. This will result in the insurance company experiencing a higher loss rate than expected. The company will then raise its rates, but this increase in price will further bias the selection of insurance towards those in a high risk class. Further iterations of this process may cause the market to disappear altogether. See George A. Akerlof, 'The Market for "Lemons": Quality, Uncertainty and the Market Mechanism' (1970), 80 *Quar. J. Econ.* 488.
136 *Taking Care of Business: Report of the House of Commons Standing Committee on Industry*, Publications Service, Parliamentary Publications Directorate, Ottawa, 1994, at 21.
137 Ibid., at 6, 17.
138 Ibid., at 17.

7

THE ROLE OF BANKS IN THE FINANCING OF KNOWLEDGE-BASED SMEs

Paul A. Toriel

CONTEXT – DEFINITION OF THE MARKET

Economies are changing rapidly, ideas and innovation are becoming increasingly the drivers of economic growth, and competition from international firms is now a fact of life for virtually all sizes of Canadian companies. It is in this environment, which has been called in Canada the 'New Economy', that Canadian firms must grow and prosper (see Table 7.1).

Small businesses, the ones with less than 100 employees, represent 99 per cent of businesses in Canada and have created most net new jobs since 1981, even in periods of recession. However 70 per cent of this growth has been created by a small number of rapidly growing businesses, about 30,000 of them, or 1.5 per cent of total small- and medium-sized enterprises (SMEs).[1] Among these 'Innovative Growth Firms'[2] leading in the fast-changing new economy, knowledge-based SMEs are playing a key role in contributing to economic growth and job creation. This is not only true for Canada, but also for the other developed countries; for example, in the United States, 4 per cent of businesses generate 70 per cent of the growth. Even in Japan, the SME sector is being looked at closely: Mr Seki, a Japanese speaker at the World conference on SMEs in Seville, Spain (September 1994) declared that 'the twenty-first century will be the century of SMEs'.

The SME sector is typically more nimble, efficient and innovative in responding quickly to market opportunities. Canadian SMEs do, however, face significant challenges in competing internationally. Until recently, one area which was a major challenge for rapidly growing, export-oriented SMEs, particularly knowledge-based firms, was access to financing.

Table 7.1 Size and scope of the New Economy (figures are for 1993 unless indicated)

Biotechnology	
World:	3,000 companies
	$15 billion in revenue
	25% yearly growth
	Forecast for year 2000: $75–150 billion in revenues
Canada:	300 companies
	13,000 employees
	$2 billion in revenue
	24% yearly growth
Medical devices	
World:	US$93 billion in revenue
	7% yearly growth
Canada:	800 companies
	17,000 employees
	$1.3 billion in revenue
	$3.3 billion in consumption
	1% yearly growth
Information Technology (includes computer industry and telecommunications)	
World:	$1.9 trillion in revenue
	Forecast for year 2000: $3.2 trillion in revenue
	Fastest growing economic sector
Canada:	15,000 firms
	Over 300,000 jobs
	$55.6 billion in revenue (1994)
	$16 billion in exports
	Most important industrial sector in Canada

Financial institutions had no choice but to innovate in order to remain major players in this new economic environment.

CHARACTERISTICS OF THE NEW ECONOMY

It is important to first provide additional information on the 'new economy' which may be categorized by the following characteristics:

1 *Knowledge intensity*: Knowledge (or information) has become a decisive factor of production, superseding land/resources, labour and capital. There is increased use and development of new technologies. We are going towards a world that will be divided between those who 'know' and those who 'know-not'. In addition to being the most important industrial sector in Canada, Information Technology has also changed the way work is being done in all other industries.

Finally, it has empowered small businesses, and they are now better able to compete with larger firms.

2 *Prominence of services*: The service sector now plays a major role in the economy and accounts for almost two-thirds of GDP. According to Statistics Canada, in 1970 the service sector represented 56 per cent of total employment; in 1994, this had increased to 67 per cent. During the same time frame, employment in the manufacturing sector decreased from 22 per cent to 15 per cent.
3 *Niche marketing*: Thanks to computers it is possible, even for small firms, to customize goods and services to meet very specific demands.
4 *Competitive advantage independent from size*: Downsizing, re-engineering and dehiring in major corporations and governments have created a shift from large to small or medium-size firms. Former executives are becoming entrepreneurs. Small companies are able to attract talent,[3] often by offering ownership. Size no longer confers the same economic advantage as in the past; large organizations often lack the flexibility required to keep up with a rapidly changing environment.
5 *Globalization*: There is a strong orientation towards exports. The distinction between domestic and international markets is eroding as trade, time and distance barriers are reduced.
6 *Continuous and rapid change*: A growing number of technologies are coming on-stream while product life cycles are shortening.
7 *Partnering*: We see the formation of new types of economic associations such as strategic alliances, joint ventures, virtual companies, and other forms of partnerships. Management is becoming more participative with a strong emphasis on teamwork; it is not unusual for key employees to be shareholders and co-decision-makers.

KNOWLEDGE-BASED SMEs

It is very difficult to specifically identify those sectors which are characteristic of the 'new economy'. An attempt to focus the chapter on 'high tech' sectors would risk missing issues related to 'low tech' sectors which have a high knowledge component in their operations. As such, I have chosen the term 'knowledge-based SMEs', defined as: firms which invest significantly in R&D to create innovation and advanced technology, and compete in global markets.

Additionally, there are a number of firms across all industries which are growth- and export-oriented and use product, process or material innovation to expand and succeed in global markets. These firms are effective users of new technologies and tend to outperform other companies in their sector. As a result, most of my comments will also apply to other innovative growth- and export-oriented SMEs which are part of the 'innovative growth firms' mentioned above.

These knowledge-based SMEs have some or all of the following characteristics: mainly intangible assets; high knowledge component to operations; highly skilled labour force; high level of R&D, high level of product/service innovation; use of advanced technology; products/services tradable internationally; and, short life expectancy for products/services. For example, some sectors which clearly would be classified as new economy sectors include computers, semiconductors, software, medical devices, biotechnology, telecommunications, advanced materials, advanced manufacturing technology, plastics processing, and the environmental industry.

Another way to divide knowledge-based firms is suggested by the National Research Council of Canada (NRC).[4] NRC divides these companies into four groups by the mode of innovation:

1 *Science-based firms*: These firms innovate through processes at the interface of basic and applied research. Example: Biotechnology.
2 *High-technology craft firms*: These depend on a continuous process of engineering innovation to build customized products. Example: Flight simulators.
3 *Systems integrators*: These buy components and integrate them into new products (e.g. telecommunications firms, aircraft builders.
4 *Flexible technology purchasers*: These are intelligent buyers of technology. Example: Some brewers and major food companies.

FINANCING ISSUES

Access to financing is a critical determinant of the success of knowledge-based SMEs. Many reports and studies in Canada and abroad have recognized the difficulties encountered by these SMEs in obtaining both debt and equity financing.

In a typical knowledge-based firm, intellectual property may constitute its chief asset and few outside the firm really understand either the business or its potential. This means it is very hard to find bankers or investors with the technical competence to be able to assess the risks involved in innovative projects, products or services. Traditional sources of financing are reluctant to provide capital secured by soft assets, and many financial institutions are unwilling or unable to develop alternative financing instruments geared to the specific needs of knowledge-based firms.

Such challenges result in a serious gap in the spectrum of equity investment and risk capital available to smaller firms. The 'fixed' cost of due diligence makes these equity investments difficult when the amounts needed in investment are under $1 million, and very difficult when the amounts needed are below $500,000.

Finally, many successfully growing companies tend to display a strong early orientation towards exporting. This requires an additional set of highly specialized skills from lenders and investors. They have to assess the exporting business against a broader range of international risk. The need for due diligence means they have to be familiar with the complexities of international trade. Since few lenders and investors possess such skills, it is extremely difficult for such SMEs to get a proper evaluation, let alone financing, for their ventures. A new breed of bankers was needed, as financial institutions were not able to keep up with the changing requirements of these new types of SMEs. We have to admit, however, many of these businesses themselves have shortcomings.

Many SMEs pursuing rapid growth are characterized by low profitability coupled with high cash requirements. For such companies, internally generated capital is insufficient to sustain growth. This increases the need for external financing while making it less likely that such financing will be available. At the same time, many SME owners and managers simply lack the skills and experience needed to manage successful growth or to develop internationally competitive firms. Furthermore, a surprising number of them do not have a clear understanding of the roles and responsibilities of the various financial suppliers. For example, many do not appreciate the different purposes served by equity and debt financing. Often they will look for debt when they need equity. And when they need equity, they will not be ready to accept to share the ownership and management of their firm with the new investor. Finally, knowledge-based SMEs until recently had few effective business networks to help them address financing. Few of these businesses are connected to larger corporations and when it comes to financing, many of these SMEs are largely on their own.

FINANCING CYCLE FOR KNOWLEDGE-BASED SMEs

It is very important to understand that the financing needs of a knowledge-based SME evolve with its growth and the financing sources are also very different. Communication breakdowns between lenders and knowledge-based SMEs are often caused by the ignorance of proper financing sources.

We can divide the growth cycle of a typical knowledge-based SME into five stages: seed, start-up, consolidation, expansion and maturity. The length of each stage will vary dramatically by industry. For example, in the biotechnology sector the seed stage may often last up to ten years, while in the software sector it may last only a few months.

During the *seed stage*, the product or service is just an idea, a concept, or a prototype; the market is unproven; and a business plan is unlikely to be developed. The entrepreneur has to provide for the initial capitalization of the firm and the first R&D related expenditures. The sources for the financing should be: first the owner (cash and 'sweat equity'); then family and friends ('love' money); and finally local, provincial or government money (It should be noted, however, that given recent federal and provincial budgets, government money is becoming increasingly scarce.)

At the *start-up stage*, we should have a proven product or service with demonstrated market potential; the firm should have made some sales or at least have firm contracts in place, and the business plan should be developed. The entrepreneur needs funds for additional capital, R&D expenditure, operational expenses, and investments in fixed assets. In addition to the sources of financing mentioned for the 'seed stage', moneys may be available now from informal private investors ('business angels'[5]), the banks and the Business Development Bank of Canada (BDC) or provincial equivalent (e.g. SDI). Near the end of this stage the company should start making profits, which will constitute another source of financing (retained earnings).

During the *consolidation stage* sales are now regular and the firm has started to export. R&D is needed to refine the product or service and ensure new ones are ready as soon as possible. In addition to all of the above, the firm needs to finance its exports. At this point, bank financing plays a very important role as well as retained earnings. The Export Development Corporation (EDC) and the Canadian Commercial Corporation (CCC) may assist in financing exports or insuring foreign accounts receivables. Additional capital, equity or subordinated loans may be needed and could be available from venture capitalists or merchant bankers.

At the *expansion stage* sales are growing very rapidly and the firm has expanded to new markets with additional products or services. New equity is needed to support higher borrowing needs. This may be available through venture capitalists, a private placement or an Initial Public Offering (IPO). If successful, the expansion stage is followed by the *maturity stage*, the last phase, in which the firm and its products or services are now established and well known. Financing should no longer be a problem and the company is often publicly traded.

RISKS FOR LENDERS, INVESTORS AND SMEs

Several types of risks affect the existence of knowledge-based SMEs and by extension their investors and lenders. The major risks to be analysed are as follows:

1 *Management risk*: This is the most important risk for knowledge-based SMEs. In most cases the solidity and competency of the management team are more important than the product or service they intend to produce. The following questions need to be answered by bankers and investors: What is the management's previous experience and qualification? Does the management have a vision supported by an achievable strategy? Are contingency plans in place? Is the management team cohesive? Will the key people stay with the firm? Are they committed?

2 *Industry risk*: The performance of the industry affects the performance of the firm. Any firm operating in a difficult industry will have more challenges to face. The following questions have to be answered: How is the industry sector performing? Is it growing, stable or declining? Is the industry segment vulnerable to a substantial technological advance in a related industry? Is the industry cyclical? What is the position of the firm within its industry (leader, outperforming, underperforming)?

3 *Market risk*: To be successful the product has to be needed and marketed properly. A lot of good products are not making it because of flaws in market research or in marketing. The following questions need to be answered: How will the new product or service be accepted by the market? How good are the estimates made by the entrepreneur regarding markets and product life cycle? Is the market broad or global enough to allow for expansion? If the sales are in foreign countries, what are the specific risks to these countries?

4 *Product/service risk*: The major concern here is the quality of the product or service. Can the company produce the product/service as advertised to buyer and at a cost sufficient to make reasonable profits? Can the product be protected by patents? Is the product environmentally sound? Is the firm following ISO 9000 criteria?

5 *Technological risk*: It is important to determine how innovative the new product is by asking the following questions: Is the company effectively managing the R&D function? Are there new products in the pipeline or do we have a 'one product company'? Are there 'quantum leaps' in technology within the sector that could create material competitive problems?

6 *Financial risk*: This is the last but not least of the major risks. Often companies do not provide for sufficient equity to leverage their present and future growth. What are the sources of capital for the future? Are there financial contingency plans in place? Is working capital managed effectively? Is liquidity adequate? Does the company need loans or equity?

All the above risks will influence the decision of the lender or investor to provide loans or capital to the company. To know these risks exist is a first step. To manage and control them is the real challenge. Often the banker/investor will help the entrepreneur to analyse the risks just by asking the above questions.

FOREIGN MODELS FOR SOLUTIONS

Two banking models for financing knowledge-based SMEs are described below, the NatWest model from the United Kingdom and the Silicon Valley Bank model from the United States.

In 1989, the National Westminster (NatWest) Bank, one of the largest United Kingdom banks, established a Technology Unit (NatWest model) within its small business service division to provide specialist support to technology-based businesses. This unit is set up as follows:

1 A small central team of ten experienced managers and research assistants supporting over 150 trained Technology Account Managers located in selected NatWest branches throughout the UK and serving local technology-based firms.
2 The Central Unit provides technology account managers with specialized training; up-to-date information; expert advice on new developments in technology and commercialization of innovative products.
3 The Central Unit actively fosters the growth of the UK technology sector by sponsoring key personnel, organizations, awards and projects as well as influencing the Bank's policy to ensure that technology clients' problems and needs are being taken into account.
4 Specialists provide practical solutions to clients, including specially designed financing products and packages, new technology appraisal services, and networking with technical and business specialists as well as 'business angels'.

This structure has allowed NatWest to dominate the technology and knowledge-based SME market in the UK while benefiting from below average loan losses in this sector and a good profitability.

The Silicon Valley Bank model was set up by the Silicon Valley Bank, a small California bank (assets of US$992 million in 1993) which emerged as a leader in lending to the technology industry with over 1,500 clients coast to coast in the technology sector (from start-ups to public companies). Their technology portfolio amounted to US$320 million in early 1994. Over the last ten years, they have experienced only US$2 million in losses in technology lending. The Silicon Valley Bank's success is attributable to the following factors:

1 Specialized experience of the lending officers operating in an integrated technology group.
2 Technology loan committee.
3 Twenty to twenty-eight borrowing accounts per loan officer.
4 Close working relationship/partnership with specialized venture capital companies to share due diligence, risk assessment, and problem resolution.
5 Outsider's scrutiny considered key factor of success.
6 Typical target client: start-up, with developed/saleable product; US$3 million and 10 million in sales (actual or forecasted); audited financials and business plan.
7 Lines of credit are essentially supported by receivables and cash flow (not geared to net worth).

Financing knowledge-based SMEs in Canada has to be considered as a long-term effort, critical to future economic growth. SMEs are seen in OECD countries as the engines for economic recovery and job creation. And among SMEs, particular attention is being paid to firms growing rapidly through innovation and exports (Innovative Growth Firms).

The following give an idea of the global importance of SMEs:

1 In the United States, small firms produce twice as many product innovations per employee as large firms. Some of the most important innovations of the twentieth century, such as the FM Radio, the microprocessor, or the optical scanner, have been made by small firms.
2 In the European Union (EU), SMEs comprise 99.9 per cent of total enterprises and provide employment for 67 million people; this represents 72 per cent of total employment in non-agricultural private enterprise. SMEs also account for 70.5 per cent of total business turnover (sales). Furthermore, SMEs are responsible for the creation of 40 per cent of all new jobs in manufacturing. This trend is strengthening and the EU economic health is dependent on a flourishing SME sector.
3 In Denmark, close to half of exports are attributable to firms with less than 500 employees. As a comparison, in Canada, only 9 per cent of exports are generated by SMEs.
4 Policies to develop SMEs, to assist in their financing, and to eliminate hindrance to their growth, are top political issues in most OECD countries. Indeed, most governments see the SME sector as generator of growth.

A PROPOSED CANADIAN MODEL: AN INNOVATIVE APPROACH

In November 1993, I was seconded, for a period of one year, by the Royal Bank of Canada to Industry Canada and was mandated to work with private and public sector providers and users of capital to develop and recommend improved approaches to the financing of export-oriented, knowledge-based SMEs.

To undertake the work, we formed a committee made up of representatives from banks; public sector financial institutions such as the Federal Business Development Bank (now BDC), the Export Development Corporation, and the Canadian Commercial Corporation; representatives from several SME associations; as well as representatives from the federal government, from two provinces and from academia. The research and analysis focused on debt and export financing. We felt that in addition to examining the domestic perspective, we had to research international best practices in other OECD countries.

To identify best international practices which could be adapted to Canada, discussions were held with representatives of government departments, industry associations, banks, export agencies, and venture capital firms in the United States and in Europe, as well as with OECD officials. Additional work involved a review of all the recent literature on financing SME issues, emanating both from Canadian and international sources. During our study, interested parties, both in the private and public sectors, were kept abreast of our progress and given the opportunity to read our research material.

In June 1994, the final report, titled *Financing the New Economy: Towards a Positive Conspiracy*, was completed and widely distributed. It contained a series of recommendations to both governments and financial institutions and a call for increased cooperation between private and public sectors. The major recommendation to banks was that they accelerate the process started by three of them to establish special units to service knowledge-based SMEs and to link these units through a special network. We felt that, based on models we had visited in England and the United States (NatWest and Silicon Valley banks), integrated networks of such 'expert' units should be established to service the needs of knowledge-based SMEs across Canada.

However, we recognized that banks had the following constraints: it would be very difficult for them to train all account managers to understand the challenges of lending to new economy SMEs and to be prepared to present them with innovative solutions; and, while knowledge-based SMEs are increasingly prevalent and located across Canada, presently they tend to be concentrated in some 20–5 areas. Therefore, our recommendations were as follows for banks considering establishing these 'expert units':

1 locating the units strategically across Canada where the need is greatest, and linking them to specialized credit officers situated centrally at the Head Office or regional office of each bank;
2 providing special training to account managers of these units and linking them across the country to accelerate the cross-training process;
3 linking the units with the banks' international trade units, since most Canadian knowledge-based SMEs tend to be exporters;
4 connecting them with the federal and provincial programs and information initiatives, and with federal and provincial agencies;
5 linking them to universities, private laboratories and/or other technology and knowledge experts where appropriate;
6 having account managers act also as pathfinders to other sources of capital, such as private investors and venture capital. (Indeed, linkages and/or active partnerships between lenders and investors are also very important to help manage the inherent risk in financing knowledge-based SMEs);
7 taking into consideration that each geographical area may have a different knowledge-based industry focus; and
8 examining internal organizational and reporting structures as well as promotion, reward and compensation programs to ensure that there are no impediments to the formation and success of these specialized units.

The emphasis is on developing a group of 'expert' account managers who can achieve long-term career goals while providing the knowledge-based SMEs with a 'stable' interface to financing. Some of the other recommendations in the Report were:

1 Banks, venture capitalists, and informal investors (business angels) should work together to share due diligence and risk assessment.
2 Banks should explore innovative ways to finance higher risk transactions. Higher pricing/equity kickers should reward their risk.
3 Banks should form alliances with other financing providers (public or private) to make available medium-term financing (mezzanine financing or quasi-equity) to knowledge-based firms with soft assets.

In the recommended model, four integrated networks (Finance, Government, Business, and Universities) are built around the knowledge-based SME. The important factor here is the cross-fertilization between the networks and the cooperation to achieve better results.

Following the lead of the Royal Bank, which has over fifty specialized account managers in more than twenty strategic locations throughout the country (see Royal Bank of Canada KBI Model, next section), other banks have established special units to service knowledge-based SMEs. These units went from three in October 1993 with a handful of specialized account managers, to more than fifty (as at July 1996) across Canada with well over one hundred specialized account managers. Bank of Montreal has eleven Innovation & Technology Centres; CIBC has some thirty specialized account managers; Toronto Dominion has eight Technology Banking Centres; National Bank has seven specialized teams; and Scotiabank has four Innovation & Growth Centres. The following results are apparent:

1 As recommended in our Report, some banks, venture capitalists and informal investors are starting to work together as partners to share the due diligence and risk assessment, and even the risk itself. Here again, the Royal Bank led the way with its 'Neuroscience Partnership'. Strategic partnerships are also being formed with 'experts' (such as research organizations, for example) who can help provide fast, reliable and cost-effective technology assessment assistance. This cooperation allows for more informed decisions on taking the risk of financing knowledge-based SMEs (and therefore is reducing the risk of losses).
2 Several banks have developed innovative ways to finance higher risk transactions (alone or in cooperation with others).
3 Venture capital firms have substantially expanded financing available to knowledge-based SMEs. In 1995,[6] $428 million, or 64 per cent of the venture capital invested, went to 221 technology companies. This activity represented an increase of 27 per cent in term of number of companies and a 58 per cent increase in the amount invested in technology companies over 1994. The average amount invested in each technology company increased from $1.6 million in 1994 to $1.9 million in 1995.

ROYAL BANK OF CANADA KNOWLEDGE-BASED INNOVATION (KBI) MODEL

Royal Bank, the largest Canadian financial institution with over 18 per cent market share of Canadian SMEs business, was the first to adopt a model inspired from the above recommendations. Three markets in which the Royal Bank had a leading edge (Information Technology, Life Sciences, Media and Entertainment) were chosen to implement the following model:

1 A network of over fifty *specialized account managers* placed in more than twenty strategic locations throughout the country. They network with each other, participate in local industry associations, get involved in local technology initiatives, and work closely with universities. Some are co-located in a specialized centre with their clients (e.g. the KBI account manager in Saskatoon is located at Innovation Place).
2 *Community Advisory Councils* located at several sites (comprised of local governments, academics and business leaders in the knowledge-based industries) act as advisers to the account managers.
3 *Specialized credit approval teams* which communicate directly with the account managers to streamline the approval process.
4 *Tailored financial products* have been developed to fulfil the special needs of knowledge-based clients.
5 *Direct links with all specialized units* of the bank (Trade Finance, Foreign Exchange, Leasing and Venture Capital) have been established. In this regard, Royal Bank also created a CDN$125 million fund to invest in knowledge-based and export-oriented SMEs.

Royal Bank of Canada was able to take a clear lead among Canadian banks in financing SMEs in the new economy by adopting the above model. Their innovative approach, centred on the knowledge-based client's needs, has attracted all other stakeholders such as venture capitalists, research institutes, universities, and government to cooperate with them in this new endeavour. Examples of the resulting new types of partnerships formed are found below:

1 ViaTech Technology Viability Network is a partnership arrangement, to provide banking, legal, accounting and marketing support to new and existing technology businesses. ViaTech forms a bridge between emerging enterprises and the established professional and service companies that can offer value to them as they grow. ViaTech groups have been formed by Royal Bank with various partners in several communities across Canada.
2 Three hundred million dollars was committed to new loan programs for growth and export-oriented small businesses. They feature flexible term loans of $50,000 to $500,000, terms of up to five years, possibility to delay principal repayment and softer collateral security requirements. They are supported by strategic alliances with regional federal government agencies. As of July 1996, the following financing programs are in place:

   - In the four western provinces, a $30 million program for biotechnology and agricultural biotechnology companies, a $20 million program for SMEs in the Health sector and a $40 million program for SMEs in Information Technology & Telecommunications; all are joint ventures between Royal Bank and WD.

- In Northern Ontario, a $10 million program to support innovative small firms in alliance with FedNor.
- In Quebec, a $50 million program for innovative small businesses in alliance with FORD(Q).

3 In March, 1996, Royal Bank formed a strategic alliance with the Business Development Bank of Canada to offer 'one-stop shopping' to innovative small businesses in Southern Ontario. This alliance offers easy access to a full range of products and services from both institutions as well as accelerated processing of joint financing for amounts from $50,000 to $1 million. This alliance is now being expanded to other locations.

4 Royal Bank has committed $1.8 million to ACF Atlantic Equity Inc., a venture capital initiative also sponsored by federal and provincial governments and other financial institutions in the Atlantic provinces. This fund, totalling $30 million, is directed to SMEs in these provinces.

5 Royal Bank has committed $500,000 to Nuvoventure Fund. This is a strategic alliance with Nuvotech Ventures International in Saskatoon, Saskatchewan (and other parties) to invest in early stage SMEs involved in agri-product development.

6 Royal Bank injected $150 million into its wholly-owned subsidiary, Royal Bank Capital Corporation. Over the past five years, RBCC has invested more than $125 million in more than a hundred growth-oriented businesses. A further $25 million has been committed through the NeuroScience Partners Fund.

EPILOGUE: THE 'POSITIVE CONSPIRACY'

In order to provide for better and more innovative financing to knowledge-based SMEs in the new economy, my report recommended in June 1994 a greater partnership between businesses, academia, industry groups, financial institutions and governments. What was only theory then was put into practice very quickly, and a considerable progress has been made in the last two years. Financial institutions and governments, as well as venture capitalists, have adopted innovative ways to deal with knowledge-based SMEs, and a new breed of bankers are now providing specialized service to these companies. Banks have become familiar with the special problems and needs of rapidly growing SMEs, and these companies are now better able to understand the role of banks.

As we get closer to the third millennium, it has become obvious that no single government or financial institution can alone solve the complex financing problems of knowledge-based SMEs. We need to continue to forge new and better partnerships between all stakeholders. This is what we call the *positive conspiracy.*

NOTES

1 See *Growing Small Businesses*, Industry Canada, February 1994.
2 This term is used by the Province of Ontario. In the USA they are called 'gazelles'; in the UK the term 'Superleague Companies' is used.
3 A recent study at Washington University found that, as a rule, it was the big firms' most talented engineers who went to the small companies.
4 See *National Systems of Innovation*, NRC, April 1994.
5 Usually business angels not only provide capital, but also become personally involved in the management of the company.
6 Canadian Venture Capital Association News Release – 3 April 1996.

BIBLIOGRAPHY

Association of Canadian Venture Capital Companies, *Venture Capital in Canada* (Annual Statistical Review and Directory, June 1994 and 1995).

Australian Manufacturing Council, *The Innovation Cycle: Practical Tips from Innovative Firms* (Melbourne: The Council, 1995).

Baldwin, John, *Strategies for Success: A Profile of Growing and Medium-sized Enterprises in Canada* (Ottawa: Statistics Canada, 1994).

Blais, Roger A. and Toulouse, Jean-Marie, 'Entrepreneurship Technologique: 21 Cas de PME à Succés', Unpublished paper, Montreal, 1992).

Canadian Bankers Association, *Small Business Annual Report* (Toronto: Canadian Bankers Association, June 1995).

Commission of the European Communities, 'Financial Problems Experienced by Small and Medium-Sized Companies' (Commission communication), Luxembourg, November 1993.

Confederation of British Industry, *Finance for Growth: Meeting the Financing Needs of Small and Medium Enterprises* (London: Confederation of British Industry, August 1993).

Doyle, D.J., *Making Technology Happen* (Ottawa: Silvan Communications, 1997).

European Observatory for SMEs (Second Annual Report, April 1994).

Industry Canada, *Building a more innovative economy (Agenda: Jobs and Growth)* (Ottawa: Industry Canada, November 1994).

Macdonald & Associates, 'The Venture Capital Market in Canada: An Analysis of 1993 Venture Capital Activity, Unpublished paper, March 1994.

Macdonald & Associates, 'The Venture Capital Market in Canada: An Analysis of 1994 Venture Capital Activity, Unpublished paper, March 1995.

MacIntosh, Jeffrey G., 'Legal and Institutional Barriers to Financing Innovative Enterprises in Canada', Unpublished draft, 1994.

Martin, Morley, *Exports and Job Creation* (Ottawa: Department of Foreign Affairs and International Trade, June 1993).

Martin, Paul *et al.*, *Growing Small Businesses* (Ottawa: Ministry of Finance, February 1994).

National Research Council, *National Systems of Innovation* (Ottawa: NRC, April 1994).

OECD, *Small and Medium-sized Enterprises: Technology and Competitiveness* (Paris: OECD, 1993).

OECD, *National Systems for Financing Innovations* (Paris: OECD, 1994).

Riding, A., Dal-Cin, P., Duxbury, L., Haines, G. and Safrata, R., 'Informal Investors in Canada: The Identification of Salient Characteristics', Unpublished paper, May 1993.

Roy, Philippe, *Vision for a Nation: Business Networks* (Ottawa: Industry Canada, June 1994).

Small Business Working Committee Report to Ministers, *Breaking Through Barriers: Forging our Future* (Ottawa: Industry Canada, December 1994).

Standeven, Peter, 'Financing the Early Stage Technology Firm in the 1990s: An International Perspective', Unpublished paper, November 1993.

Statistics Canada, *Research Paper Series: Nos. 71, 73, 74, 76, and 96* (Ottawa: Statistics Canada).

Toriel, Paul A., *Financing the New Economy: Towards a Positive Conspiracy* (Ottawa: Industry Canada, June 1994).

8

COMMENTARY

Larry Wynant

Professor MacIntosh's chapter does an excellent job of synthesizing conceptual considerations and empirical research evidence to predict the role banks can play in financing high technology SMEs. The chapter also meshes nicely with the issues Paul Toriel raises in discussing the approaches used by Canadian banks, adding one critical element that is not discussed in Toriel's chapter: the importance of collateral in banks maintaining a low risk lending policy.

MacIntosh correctly points out that banks are in the low risk lending business. (Our last study concluded that banks can even be considered to be, a priori, in the *no risk* lending business, though clearly not so in an *ex post* sense.) A low risk lending posture is achieved through two key vehicles:

1 a careful selection of clients who demonstrate very strong prospects of success; and
2 the availability of collateral to provide for a back door means of recovering outstanding loans.

Therefore, the absence of tangible, non-'transaction specific' assets will be a major hurdle to banks providing more support for high technology businesses.

MacIntosh carefully examines a series of factors which suggest SMEs, and high technology firms in particular, will be viewed as high risk businesses by bankers:

1 the absence of bankable assets (because assets are not tangible or are very transaction specific);
2 significant levels of systematic risk, largely because of higher than normal levels of leverage (however, in general, there is no reason to believe that SMEs will be engaged in businesses with inherently higher than average levels of risk); and

3 higher levels of unsystematic risk (since SMEs are less diversified), although MacIntosh argues that this risk should be of little concern to banks since it is diversifiable.

More research is necessary to determine if diversification of loan portfolios can eliminate unsystematic risk when there is a tight ceiling on loan returns combined with the chance of a 100 per cent loss if the business fails.

In addition, as MacIntosh points out, bankers do not act as if systematic risk is the only risk element that matters. Other SME risk elements include:

1 risks associated with a one-person organization;
2 higher lending costs (selection, loan negotiation and documentation, monitoring, collateral realization, and reorganization costs); and
3 higher agency costs (risk shifting, asset stripping, underinvestment, trumping debtholders' claims, shirking, incompetence) resulting from the nature of SMEs, asymmetric information and the general absence of market controls that larger firms experience.

Two aspects of lending to SMEs, in particular, are worthy of note. First, MacIntosh concludes that risk shifting may be less severe for SMEs because of the less diversified asset holdings of the owner/operator; however, our last study found that many SME managers greatly underestimate the risk facing their business (two-thirds feel their business is less risky than the average in their industry or market). Second, the ability of bankers to understand the high technology of SME clients is low due to the complexity of such businesses, the high workloads for account managers, and a general reluctance by owners/operators to share fully the details of their business with bankers.

MacIntosh proceeds to examine the likely response by bankers to higher than average risk borrowing prospects: collateralizing the loan, securing personal guarantees, screening loans, imposing debt covenants, shortening the terms of loans, taking equity participation, and charging higher interest rates. He concludes that many of these mechanisms are difficult to achieve with SMEs. Therefore, bankers will be more selective ('ration') in lending to SMEs and high tech businesses.

MacIntosh then examines empirical evidence to determine if high tech SMEs do, in fact, have higher turndown rates or pay higher loan costs. He points out, quite appropriately, that the available evidence on turndowns (largely the CFIB survey) is seriously flawed. The same holds true for borrowing cost comparisons, as loan application and management fees are now very significant components of lending costs. I cannot agree, however, with his conclusion that SMEs are undercharged by the banks because of political pressures. The availability of data on the profitability of various categories of commercial borrowers over time is insufficient to arrive at any definite conclusions. Nevertheless, my consulting work with banks leads me to suspect the view that banks undercharge SMEs.

MacIntosh's chapter raises two critical issues that the banks and the government sector need to resolve:

1 *Can the role of banks be expanded to take on higher risk lending?* The dominance of the banks in the SME marketplace naturally causes the government sector to look to banks as the provider of a broader range of financing support. However, there is no evidence that there is a capital gap for SMEs; in fact, the work by Alan Riding and his colleagues suggests that there may be more than adequate sources of higher risk financing available. The challenge is to increase the willingness and capacity of SMEs to access these sources.

   The banks are now specialized in low risk lending and have the benefit of several centuries' experience in developing loan granting and management policies that permit them to be efficient in this task. The evidence seems strong that banks, in fact, provide very low cost financing to the businesses that fall in their market set. As MacIntosh correctly points out, changing the banks' speciality will not be a simple task.

2 *Is there hope for more low cost bank financing to be provided to high technology SMEs?* MacIntosh concludes that this role will be limited, particularly for early stage businesses. However, there are other parallels within the industry that suggest the banks can overcome the two key obstacles he raises, understanding high tech businesses sufficiently well to develop a comfort level in assessing their risks and creating collateral vehicles.

There have been other instances where the banks have been successful in this task – project financing in the natural resource and extractive industries is one such example. In this case, the banks recruited and trained a specialized set of account managers, and then developed new techniques for appraising risk and securing bank obligations. The same approach is necessary for high tech SMEs, and it is to this challenge that Paul Toriel's chapter provides useful insights.

Paul Toriel's chapter describes a major cooperative effort in the banking and government sectors that is unique in terms of its speed of problem identification, the inclusion of a wide range of groups essential to the financing success of SMEs, and the recommendation of clear initiatives to deal with the problems identified. The 'SME Financing Cycle' in Toriel's chapter is an extremely useful way of describing the role of various funds suppliers to SMEs, and their interactions. The cooperative study described in the chapter and the subsequent banking initiatives are to be commended.

Actions through government initiatives are also necessary; however, stimulating new government programs will likely require further definition of the knowledge-based sector and clarification of its importance to Canada, such as:

1 the definition of a 'knowledge-based' business requires more precision to ensure that financial institutions, governments and public sector institutions can really focus their efforts on the 'right' businesses;
2 'knowledge intensity' needs to be defined in a useful manner;
3 the issue of whether 'low tech' businesses are to be considered part of the knowledge-based sector needs to be resolved;
4 the importance of exporting for these businesses needs to be determined (SMEs account for only 9 per cent of exports); and
5 more data is needed to confirm that this sector is worthy of special focus by the government, including information related to the importance of knowledge-based SMEs in job creation and economic growth, determination of whether the long run financial performance of these firms is strong enough to ensure they will be attractive to financial institutions (the Silicon Valley Bank knowledge-based SME portfolio of US$320 million is too small to be of much interest to our banks with assets in excess of CDN$150 billion), and evidence of whether nimbleness will actually supplant size as a key competitive factor.

Another critical issue raised by Toriel is why it has taken Canada so long to adapt its export financing services to meet the needs of our business sector. Will the initiatives described in his chapter (an exporter's road map, new EDC guarantees, seminars) be enough, given that these problems persist in spite of a country which leads its G-7 partners in terms of the importance of exports to GDP (30 per cent)?

The approaches being used by banks to deal with knowledge-based SME customers has great appeal to the handling of other SMEs, and one hopes that the creation of specialized KBI groups within the banks is the forerunner of a more pervasive banking reorganization in future years. Our 1991 study of banks and small business borrowers found a widespread incidence of the problems mentioned by Toriel for KBIs:

1 business owners/operators who have a poor understanding of what banks and other financial institutions really do and limited skills to interact effectively with them;
2 bankers who have a very limited understanding of their customers' businesses because of high turnover, large client loads (100+), and the need to service clients from a broad range of industries;

3 owners/operators who lack the skills to understand and manage their financing needs and performance;
4 SMEs which are unwilling or unable to communicate accurately their needs and prospects; and
5 bankers who are ill-informed or poorly motivated to recommend other sources of financing for clients that do not warrant bank support.

What is unique about the approach being used by several banks is the creation of focused banking groups, smaller account loadings, new financing products, deliberate linking of the various experts within a bank, and unions between banks and other sources of financing support. Whether this approach can be used for other SME sectors probably depends on the profitability banks can achieve with knowledge-based clients, which will take some time to demonstrate.

Finally, with respect to Toriel's 'positive conspiracy' approach, past research suggests that the banks will face particular challenges in achieving success, including:

1 the need for accounting bodies to modify the criteria for identifying and accounting for value in businesses;
2 the fact that the skills of knowledge-based SME managers will be critically lacking in business planning and managing financial needs and performance, due to the fact that an intensity of technical knowledge skills means (by definition) a dearth of business skills and that a lack of planning and financial management skills will, I believe, constitute the most critical element of management risk that bankers will have to assess;
3 the fact that recruiting and selection of KBI bankers and their training will be a particular problem;
4 the difficulty of expecting bankers to stay on top of industries where 'quantum leaps' in technology are possible (expected), initially only understood by the few firms successful in such a process; and
5 the difficulty faced by business schools, that have traditionally generated bankers, in adapting their curriculum to produce bankers capable of managing the technical as well as financial aspects of client businesses which are subject to 'quantum leaps' in technology.

Part IV

COMPETITION AS COOPERATION

Law and economics perspective of cooperative industrial policy

9

COMPETITION POLICY AND COOPERATIVE INNOVATION

Donald G. McFetridge

INTRODUCTION

The purpose of this chapter is to examine the role of public policy in facilitating and encouraging the development of national innovation systems. Students of the innovative process see it as inherently interactive. The institutional arrangements for managing this interaction comprise an innovation system and are continually evolving. In recent years, much attention has been paid to strategic alliances and R&D or technology consortia.

There have been numerous attempts to categorize strategic alliances and to distinguish between strategic alliances and consortia (Kumar and Magun, 1995). The institutional details do differ and the differences do matter in so far as they are reflective of the technological objectives of the participants and the contracting hazards which confront them. Whatever their institutional form, however, these arrangements have the common fundamental motive of exploiting knowledge complementarities.

Government policy can accommodate and sometimes encourage cooperative innovation. For reasons given below, the accommodation role is likely to be much more important than the encouragement role. Competition and regulatory policy are central to the accommodation role of government. In the past, neither competition nor regulatory policy have been particularly well informed by an understanding of the innovative process. The purpose of this chapter is to assist in changing this situation.

GENERAL CHARACTERISTICS OF STRATEGIC ALLIANCES

Strategic alliances have been defined by Khemani and Waverman (1992) as:

> [a] form of inter-firm agreements or arrangements between independent firms which involve knowledge production or sharing activities aimed at developing new products or processes and new forms of production. In this regard, the alliance may entail exchange of R&D and or transfer of various information.
>
> (Khemani and Waverman, 1992: 4)

Strategic alliances have many definitions but there is some consensus regarding their common characteristics. They are arrangements between independent firms (i.e. firms that are not jointly controlled). They are entered into for the purpose of new product, process or market development. They involve the exchange of complementary intangible assets. Intangible assets are information in various forms including tacit and codified knowledge and reputation.

Strategic alliances generally involve reciprocal exchange or swaps. There are several reasons for adopting this type of arrangement. It reduces the need to price these assets. Instead, they are traded for something that is perceived to be of equivalent value. This facilitates trade in the core assets of a firm. Core assets are central to a firm's market identity or competitiveness and, as a consequence, are especially difficult to price.

Reciprocity reduces the considerable monitoring and enforcement costs normally associated with transactions in information. Each party reveals something of roughly equal value to the other. The threat of the loss of reciprocal benefits reduces the incentive for opportunistic behaviour (Williamson, 1983). Each party in an alliance can credibly threaten to punish a lack of cooperation by the other party with a reduction in its own contribution. An important virtue of strategic alliances is that they avoid the detailed specification of obligations, procedures and limitations.

CHARACTERISTICS OF CANADIAN STRATEGIC ALLIANCES

A recent study by Magun (1996) has thrown some light on the nature and objectives of strategic alliances entered into by Canadian firms. Magun finds that the most important motive for entering a strategic alliance is to gain access to new markets, often foreign markets.[1] Consistent with this is the finding by Magun that distributors are the most common type of partner in a strategic alliance (35 per cent of all alliances). Thus, the most frequently observed strategic alliances in Canada are vertical and market development oriented.

Other motives for strategic alliances (in order of importance) include: gaining access to new technologies or resources; reducing financial risks; integrating markets and technologies; speeding new product development;

reducing R&D risks or coping with escalating R&D and technology costs; attaining cost competitiveness. The importance of knowledge acquisition as a motive for entering a strategic alliance is apparent from this list.

Other types of partners in strategic alliances include competitors (20 per cent of alliances) and suppliers (13 per cent of alliances). The remaining 32 per cent of alliances are with firms that are neither competitors nor distributors nor suppliers. The vast bulk of alliances (80 per cent) apparently do not involve firms that regard themselves as competitors and may thus not pose any threat to competition.

IMPLICATIONS OF DEVELOPMENTS IN INDUSTRIAL POLICY FOR STRATEGIC ALLIANCES

Industrial policy has historically taken what could be termed a Pigouvian approach. This involved the use of taxes, tariffs, subsidies and procurement to alter the composition of national economic activity. Certain firms, industries and regions were favoured over others. In its most sophisticated form, this involved the use of tax and subsidy measures to compensate for various perceived market failures. This could be termed 'first best' industrial policy. An example might be a system of subsidies and tax incentives for scientific research and development calibrated so as to compensate for the inappropriability of the benefits of R&D.

A 'second best' form of industrial policy which has been much discussed in recent years attempts to exploit market imperfections to benefit the domestic economy at the expense of foreigners. This is called strategic industrial policy. It has its roots in the optimal tariff literature which recognized that a country with market power could use a tariff to improve its terms of trade. This idea was expanded in the strategic industrial policy literature of the 1980s to include a variety of tax, subsidy and market-closing measures that governments might theoretically take to shift oligopoly profits from foreign firms to domestic firms. Economists have been sceptical of the practical relevance of this theoretical literature. Among the leading sceptics are some of the leading contributors to this literature (Grossman, 1992; Krugman, 1994).

In practice, industrial policy has frequently proven to be just one more manifestation of rent-seeking (Trebilcock *et al.*, 1990). The redistribution it brought about was largely within the domestic economy (Harris, 1989). Indeed, rent-seekers and their lobbyists have become adept at wrapping themselves in the flag of market failure. This approach to industrial policy is slowly being abandoned. There are three reasons for this. First, there has been some learning. The tax-subsidy approach has been seriously discredited, and writers on both business strategy (Porter, 1991) and economics (Nelson,

1992) have emphasized in its place the development of institutional linkages, whether in the form of clusters, networks or innovation systems. Second, the scope for domestic subsidy programs is increasingly limited by international trade agreements and remaining subsidies are vulnerable to countervail. The third and probably most important reason is that governments no longer have the free cash flow to finance industrial support schemes.

As it has evolved, industrial policy now focuses more on institutional factors. More precisely, it is increasingly being focused on intra- and inter-institutional relationships and the determinants, especially the public policy determinants, of their characteristics. The web of linkages among a nation's technological and financial institutions has become known as its technological and financial infrastructure. Industrial policy is increasingly oriented towards encouraging the development of this infrastructure. This approach might be termed 'Coasian' in nature. More cynically, it has been called 'industrial policy without money'.

The new approach to industrial policy can be seen in the changing program objectives of Industry Canada and the criteria of other granting agencies. It can also be seen in a number of provincial initiatives and in the changing mandates and activities of the federal laboratories which are taking on an increasingly collaborative and commercial orientation (Niosi and Manseau, 1994).

The mandate and philosophy of the Department of Industry, Science and Technology (now Industry Canada) began to change in 1988 and this process continues. Grants are being replaced by repayable loans.[2] Support is focused more on 'pre-competitive' or 'pre-production' activities such as feasibility studies, technology acquisition, R&D and modernization and on the provision of so-called 'value-added services' to firms by the department. Support is either not targeted or is targeted on broad classes of products such as microelectronic components and systems, advanced materials or biotechnology. Program criteria emphasize inter-firm cooperation, alliances and networking. Alliances may include foreign firms or non-profit institutions. In sum, the industry support programs of Industry Canada are being configured to be broadly available, to afford national treatment to foreign firms, to tend towards infrastructure rather than production and to contain a smaller element of explicit subsidy.

Some of Ontario's initiatives also reflect a concern with financial and technological infrastructure. For example, the government and six private sector partners contributed $70 million to a fund to be invested by 'expert investment corporations' (EICs) formed by investment managers, venture capitalists and others. A further contribution was expected from the EICs themselves. In order to provide a financial incentive for private sector participation the government limited its participation in any profits on investments made by the fund (*Research Money*, 24 November 1993: 2). Through its Network Infrastructure Program, Ontario also supported the construction of a fibre-optic telecommunications network linking six telecommunications companies, two universities and two federal laboratories in the Ottawa area.

Similar developments have occurred in Quebec. Innovatech, for example, is a high-tech investment fund specializing in the greater Montreal area. This fund is administered by an independent board of directors drawn from industry and operates in cooperation with other investment funds. Much of its portfolio is devoted to equity in start-up ventures. It is said to operate with a high level of expertise and with a minimum of bureaucratic interference (*Research Money*, 28 April 1993: 1–2; 24 November 1993: 3). Quebec is also promoting industry–university linkages.

While governments may continue some direct promotion of collaboration, their most important role in the future will be as facilitators. This involves ensuring that public policy accommodates rather than impedes economically beneficial forms of cooperation. This will continue to involve expenditure policies to a degree. For example, government activities can be located so as to contribute to the realization of agglomeration economies. Porter (1991) has noted that past locational decisions frequently did the opposite.

The legal framework in general, and competition and intellectual property law in particular, will be subject to increasing scrutiny and analysis from an industrial policy perspective. Examples include the analysis of the effects of patent breadth, novelty requirements and the enablement and equivalents doctrines on the rate of technological change (Scotchmer, 1991; Merges and Nelson, 1990), and the analysis of 'industrial policy' defences in merger cases (Pitofsky, 1992). These will be important in that they will determine the extent of innovative interaction and whether efficient collaborative arrangements are chosen.

The legal and administrative framework will also be important in another way. National governments have an incentive to structure this framework so as to advantage domestic firms at the expense of foreigners without violating the national treatment provisions of international agreements. The Japanese patent system and trade secrecy law are cases in point (Ordover, 1991). The same might be said about Japanese enforcement of its anti-monopoly law. Negotiations to establish a harmonized international legal and administrative framework are likely to be approached by national governments in the same way.

DEVELOPMENTS IN THE THEORY OF TECHNOLOGICAL CHANGE

Contemporary models of technological change emphasizes that it is continuous, cumulative and interactive. It proceeds in successive small, frequently unremarkable steps. Each step builds on the previous steps. Each step draws on its predecessors but also renders them obsolete. Preceding steps are both complements to and substitutes for their successors.

The so-called *feedback model* of technological change emphasizes that innovation is an interactive process. The participants in the innovative process benefit from each other. Innovation does not necessarily proceed in a linear progression from basic to applied research and then to development and commercialization.

Basic science does not proceed autonomously. It is responsive to economic incentives. It is influenced by insights arising from applied research. It can be facilitated by technological improvements in scientific instruments and calibration and measurement methods, or in manufacturing techniques. Developments in basic science need not precede technological innovation. There are many cases (metallurgy and petroleum refining, for example) in which technological innovation occurred in advance of the understanding of the basic scientific principles underlying it.[3] Moreover, the scientific knowledge on which many technological innovations are based is often drawn from developments in distant fields or disciplines.[4]

Users of new technologies actively participate in the innovative process. In the stylized linear model, users are simply at the end of the pipe accepting whatever comes out. Rosenberg (1982) provided examples of cases in which users were responsible for ideas leading to innovations and participated in the innovative process itself. More pervasive, however, is the participation by users in the cumulative process of making small improvements in existing technologies such as the steam engine or the passenger jet. Rosenberg called this process *learning by using*. He emphasized both its collaborative nature and its quantitative importance. It is the accumulation of small modifications and improvements that is responsible for most of the productivity gain resulting from the introduction of a new technology.

Many have elaborated on Rosenberg's work. New issues have also emerged. The linkage between science and technology is becoming closer. Computer-assisted research and engineering has compressed development periods. Scientific discoveries have immediate commercial applications more frequently than before. Engineering is becoming increasingly science-based. Technological innovations are more likely to have immediate scientific relevance. The distinctions both between science and technology and among the stages of the innovation process are blurred.

Paul David (1993) maintains that research and development personnel are no longer distinguished on the basis of what they do but rather on the basis of the economic incentive system under which they work. The categories of basic and applied research and development are no longer useful. A more useful distinction is between open access and proprietary (restricted access) knowledge.

Knowledge may be codified or tacit. Codified knowledge is more readily accessible than tacit knowledge. What is codified and what remains tacit is, in part, a matter of economic incentives and in part a matter of technological possibilities. Knowledge may be kept in tacit form to increase its appropriability; that is, to keep it out of the stock of open science.

It is also argued that the increasing use of computer-aided design, experimentation and testing facilitates formalization and codification of an increasing portion of the knowledge stock. An implication is that knowledge is becoming increasingly transferable. Diffusion is becoming faster and less costly. Secrecy is becoming a less viable option and the degree of appropriability may be declining as a consequence.

Successful innovations more frequently involve recombination or integration of existing knowledge from diverse sources. Within existing innovative organizations there is increasing emphasis on access to 'outside' sources of knowledge. Applied research and development is valued for its 'swap' value in a strategic alliance as well as for its internal commercial potential. Research, including basic research, is valued for the insight it gives those involved into the possible applications of the ideas of others.

The change in the nature of the innovative process has been accompanied by many institutional changes. The linear concept of in-house R&D followed by commercialization is an increasingly inappropriate characterization of the innovative process. Indeed, the period of dominance of the innovative process by the corporate R&D lab may be coming to an end and is being superseded by 'virtual R&D labs' involving networks, alliances and other cooperative arrangements among individuals ('lone wolves') and academic and government research institutions as well as firms.

IMPLICATIONS OF THE FEEDBACK MODEL OF TECHNOLOGICAL CHANGE FOR PUBLIC POLICY TOWARDS STRATEGIC ALLIANCES

Innovation as a cumulative process

The model of technological change described above sees the process as cumulative, continuous and interactive. As a cumulative process, innovation builds on the past. The current generation of innovators benefits from the wisdom of their predecessors but also renders them obsolete. The innovative process necessarily involves the sharing of information between originating innovators and follow-on innovators who are their potential competitors.[5]

Moreover, if it is to be organized efficiently, the economic incentives for engaging in innovative activity must reward successive contributions according to what they teach, enable or facilitate. In situations where one innovation builds on another but also renders it obsolete, potential originating and follow-on inventors will have to reach an agreement on the distribution of the surplus resulting from their respective efforts if either invention is to occur.

This is illustrated in Figure 9.1. In this figure, the open access technology allows its user to produce at unit cost c_0. An initial process innovation reduces unit costs to c_1, yielding a maximum of \$A per period in rents to the innovator. This innovation makes a further innovation possible. This results in a further reduction in unit costs to c_2. The follow-on innovator can undercut the initial innovator and earn rents equal to \$B per period, which reduces the initial innovator's profits to zero. Anticipating this possibility, the first innovator may not commit resources to innovation, with the result that neither innovation occurs. An agreement to suppress competition between the initial and follow-on innovations increases surplus by \$A + \$B relative to the situation in which neither innovation occurs. That is, consumers are no worse off and innovators are better off by the amount \$A + \$B per period than if the innovation had not occurred. Of course, if both innovations would have occurred without this agreement then its effect is to reduce aggregate surplus by \$L per period, which is the amount of the deadweight loss resulting from a price of c_0 rather than c_2.

The key question here is whether the sequence of innovations would have occurred if the original innovator had anticipated unrestricted competition from follow-on innovators. Where this is uncertain, it is preferable for the competition authorities to take a relatively liberal approach. Prohibition of an agreement that is necessary for an innovative sequence to

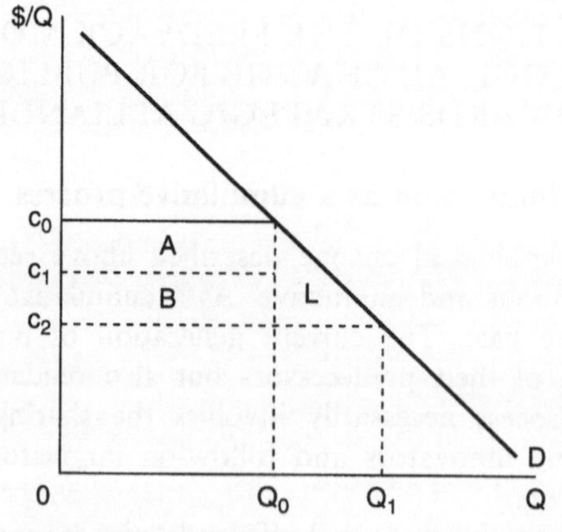

Figure 9.1 Sequential process innovations.

occur (a Type I error) results in losses to society of \$A + \$B per period. Failure to prohibit an agreement that is not necessary for a sequence of innovations to occur (a Type II error) results in losses to society of \$L per period. The cost to society of mistakenly prohibiting an agreement that is necessary for a sequence of innovations to occur is likely to be much greater than the cost of failing to prohibit an unnecessary agreement.[6]

Agreements between prior and subsequent innovators may be made before resources have been committed to either stage. These might be termed *ex ante* agreements. Agreements may also be consummated after an initial innovation has occurred but before resources have been committed to follow-on innovations. Agreements may also be reached after both prior and subsequent innovations have occurred. The characteristics of these agreements will depend, in part, on the stage in the innovative sequence at which they occur. To the extent that they have any discretion over it, innovators will time their agreements so as to maximize their joint surplus.

Agreements may take many broad forms including mergers, R&D joint ventures, R&D consortia, patent pooling, cross-licensing or licensing, as well as the type of open-ended information sharing that has come to be known as a strategic alliance. The form these agreements take depends in part upon their timing. Thus, R&D joint ventures are frequently defined as agreements entered into in the early stages of the innovative sequence, while cross-licensing is associated with the later stages of the sequence. While they may differ in timing and thus in form, these agreements may be very similar in their objectives. From a public policy perspective, it is important that agreements between or among innovators be evaluated on the basis of their objectives rather than their timing and form.

The timing and characteristics of agreements between prior and subsequent inventors depend on intellectual property law, trade secrecy law and competition law. There is both a scale effect and a substitution effect here. That is, the policy environment may be so unpromising that an innovative sequence simply does not begin. The sequences that do occur may be diverted, slowed or accelerated by the policy environment. This is especially likely to occur if public policy is biased in favour of some forms of agreements over others. Bias could arise if agreements are evaluated on the basis of their timing and form rather than on their objectives.

Competition policy plays a more important role than is commonly recognized. It determines the extent of the exclusive right embodied in intellectual property. In the past, particularly in the United States, competition or antitrust policy has paid little regard to its effect on innovation. For example, a patentee's exclusive right is effectively limited to that of a simple monopolist. Tying, bundling, forcing and other forms of discriminatory pricing are generally forbidden. The same holds true for the owners of trade secrets (Miller, 1989). Moreover, restrictions on licensing arrangements are such that a patentee relying on licensing would not likely be able to earn even simple monopoly profits.

An intellectual property owner's exclusive right may be similarly circumscribed in Canada. Under Section 32 of the Competition Act, licensing arrangements that are found to lessen competition unduly can be declared void, or their execution enjoined in whole or in part. There are no reported cases under Section 32. Two cases brought under earlier legislation were settled.[7] These cases, both of which involved provisions in licences issued by Union Carbide for the use of its patented processes and machines to extract polyethylene film from resin (the extrusion patents) and to treat this film for printing (the printing patents), illustrate the problems that competition policy can pose for efficient licensing. With regard to the printing patents, the licensing practices at issue were: royalty rates that decreased as volume increased; field of use restrictions; no challenge clauses; post-expiry restrictions on licensees. The practices at issue with respect to the licensing of the extrusion patents were the imposition of higher royalties on licensees not purchasing resin from the patentee or his nominee and post-expiry restrictions on licensees.

The apparent concerns of the Director of Investigation in these cases were with discrimination against low volume Canadian licensees, with the limitation of export opportunities for Canadian licensees and with the resin tying arrangement. While the terms and conditions of these licences may have been disadvantageous to some or all of the Canadian licensees, the licensing practices listed by the Director are potentially surplus-increasing when viewed from a global perspective.[8]

To summarize, agreements between or among successive innovators may be more similar in purpose than they are in form. These agreements may support the exchange or transfer of knowledge. They may also limit competition between the originating and follow-on inventions. In some cases this will be necessary in order to provide an incentive for a sequence of innovations to occur.[9] Thus, agreements between prior and subsequent innovators may provide for both an explicit lessening of competition in the form of cross-licensing, tie-outs or territorial exclusions as well as the exchange of information. The exchange of information may itself support a tacit agreement to lessen competition and thus be a facilitating device. Both the tacit and the explicit restriction of competition should be evaluated in the context of the cumulative innovative process. That is, the effect of the agreement on the incentive to innovate should be taken into account.

Innovation as an interactive process

As a continuous, interactive process, innovation is not confined to formal R&D. It may involve functions such as production, logistics and distribution. It may involve suppliers, customers, competitors and potential competitors. It is ongoing. It may involve many small, unremarkable improvements.

This has two very important implications. First, the distinction between R&D and other functions may not be a meaningful one. Rules that attempt to confine cooperation or information exchange to specific R&D projects are likely to deter a significant amount of innovative activity. Second, the concept of pre-competitive R&D may be similarly vacuous. Innovation may occur during or after commercialization as well as before. There is no reason to expect that the innovative process is typically characterized by a product cycle with well-defined segments in which there is innovation and other segments in which there is not. Rules that attempt to define a specific stage, such as development or commercialization, at which information exchange should cease are likely to deter a significant amount of innovative activity.

The feedback model also implies that innovation may involve co-operation in R&D and in 'production'. Confining cooperation to the research stage may be difficult. In particular, it may be difficult to distinguish anti-competitive agreements on the basis that they involve the sharing of production information. The concept of pre-competitive research may no longer have much meaning.

Strategic alliances frequently involve swaps of information. The implication is that if participation in an alliance is contingent on having knowledge to trade, firms with aspirations but no knowledge to offer could well be excluded. Thus, the form of strategic alliances is such that they may exclude potential competitors and reduce future competition. The potential losses that may ensue must be balanced against the virtues of the swap arrangement itself. This exercise will be difficult as it requires trading-off current savings in transactions cost (which are themselves not readily measurable) against a reduction in competition in a product market that does not presently exist.

To summarize, strategic alliances frequently involve the exchange of information between competitors and potential competitors. Strategic alliances are unlikely to be confined either to formal R&D activities or the 'pre-competitive' phase of the product cycle. It will not be simple to identify the information exchanges that serve principally as facilitating devices. Strategic alliances may also contain explicit provisions to lessen competition and may also appear exclusionary. Whether strategic alliances directly lessen or merely facilitate the lessening of competition, they should be evaluated from a precedential rather than a static perspective (Lewis and Yao, 1995). That is, alliances should be evaluated from the perspective of whether the sequence of innovations flowing from them would likely have occurred or would have occurred in as timely a fashion if competition had not been restricted.

APPLICATION TO COMPETITION POLICY

The US approach: The Department of Justice–Federal Trade Commission *Antitrust Guidelines for the Licensing of Intellectual Property*

The 1995 *Antitrust Guidelines for the Licensing of Intellectual Property* describe the general approach which US antitrust agencies take or propose to take in dealing with cooperative innovation. The *Guidelines* show a new awareness of the link between both licensing practices in particular and inter-firm arrangements in general and the incentive to engage in innovative activity. Section 2.3 of the *Guidelines* states:

> Licensing, cross-licensing, or otherwise transferring intellectual property . . . can facilitate integration of the licensed property with complementary factors of production. This integration can lead to more efficient exploitation of the intellectual property benefiting consumers through the reduction of costs and the introduction of new products. Such arrangements increase the value of intellectual property to consumers and to the developers of the technology. By potentially increasing the expected returns from intellectual property, licensing also can increase the incentive for its creation and thus promote greater investment in research and development.
>
> Sometimes the use of one item of intellectual property requires access to another. An item of intellectual property 'blocks' another when the second cannot be practised without using the first. For example, an improvement on a patented machine can be blocked by the patent on the machine. Licensing may promote the coordinated development of technologies that are in a blocking relationship.
>
> Field-of use, territorial and other limitations on intellectual property licenses may serve pro-competitive ends by allowing the licensor to exploit its property as efficiently and effectively as possible. These various forms of exclusivity can be used to give a licensee an incentive to invest in the commercialization and distribution of products embodying the licensed intellectual property and to develop additional applications for the licensed property. The restrictions may do so, for example, by protecting the licensee against free-riding on the licensee's investments by other licensees or by the licensor. They may also increase the licensor's incentive to license, for example, by protecting the licensor from competition in the licensor's own technology in a market niche that it prefers to keep for itself. These benefits of licensing restrictions

apply to patent, copyright, and trade secret licenses and to know-how agreements.

While they recognize the role of intellectual property agreements in promoting innovation, the antitrust agencies also view these agreements as having the potential to reduce competition between rival technologies, facilitate collusion and foreclose technologies from use by potential competitors. In their assessment of the competitive impact of an agreement the antitrust agencies have indicated that they will examine not only the conventional product market but also the 'technology market' and the 'innovation market'. The technology market is defined to include the technology that is the subject of the agreement and any close substitutes for it. The innovation market is defined to include research and development directed to particular new or improved goods or processes and close substitutes for that research and development (section 3.2.3). An agreement may run into trouble if it lessens competition substantially in any one of these markets.

In their discussion of the manner in which the antitrust agencies might define an innovation market, Gilbert and Sunshine (1995) imply that the agencies would regard any two firms with the same broad technological capabilities as being able to innovate independently and would thus be suspicious of arrangements between them unless there were at least four (section 3.2, Example 4 in the *Guidelines*) other firms with closely substitutable capabilities. This is troubling for a number of reasons. First, alliances are likely to involve firms with similar broad capabilities. Participants differ, however, in their specific capabilities. Alliances exploit complementarities in these specific capabilities. Second, whether both firms have the technical capability of proceeding independently is beside the point. The relevant questions are whether they *would* proceed independently given the anticipation of competition from the other and whether they *should* proceed independently; that is, whether more technological rivalry is necessarily preferred to less.

The *Guidelines* state that individual provisions of licensing agreements will be subject to rule of reason analysis unless they lessen competition without offering the promise of offsetting efficiencies. In this, the *Guidelines* confirm current US practice with respect to horizontal restraints. In general, restrictive provisions are regarded as potentially efficiency enhancing if they are reasonably necessary to 'facilitate the combination of the licensor's intellectual property with complementary factors of production owned by the licensee' (section 3.4). The trade-off is qualitative and differs according to the market context.

To summarize, the *Antitrust Guidelines for the Licensing of Intellectual Property* recognize correctly that so-called restrictive provisions in licensing arrangements may serve to encourage diffusion as well as innovation, and that agreements which restrict competition in the product market should be assessed from a dynamic as well as a static perspective. While the adoption of a longer term perspective is welcome, it does not yet appear to have been recognized that agreements which reduce the amount of resources devoted to innovation and even slow down the process can be socially beneficial under some circumstances.

The Canadian approach: *Strategic Alliances Under the Competition Act*

In his policy statement *Strategic Alliances Under the Competition Act*, the Director of Investigation and Research (1995) attempts to clarify the manner in which various sections of the Competition Act would apply to collaborative arrangements between or among firms in Canada. The section of the Act the Director would apply will depend on the form of the arrangement. Thus, a strategic alliance could be dealt with under either the civil or the criminal provisions of the Act. The evaluation standard and the defences allowed depend on the section of the Act applied.

The discussion in the policy statement suggests that horizontal alliances posing potential competition problems are most likely to be dealt with under section 45. Under section 45, agreements which, if carried out, would have the effect of lessening competition unduly are a criminal offence. There is no provision under section 45 for consideration of efficiencies or other offsetting benefits flowing from an agreement. Even if the jurisprudence under section 45 allowed them to look beyond the public interest in free competition, the criminal courts are, as Kennish (1994) has noted, poorly suited to do so.

Agreements relating to R&D are exempt under section 45(3) provided they do not have the effect of preventing or lessening competition substantially with respect to prices, product quantity or quality, markets or channels of distribution. While the scope of this exemption remains to be tested in the courts, there would not appear to be any defence available for an R&D or technology-sharing agreement that has been found to lessen competition unduly with respect to one or more of prices, quantities, quality, markets or channels of distribution. Thus, a cross-licensing agreement which lessened competition unduly in one or more of these dimensions could not be defended on the grounds that it eliminated duplicative R&D or allowed for more efficient timing of R&D or avoided litigation costs.

Indeed, the fact that an agreement is confined to R&D and does not have provisions (market allocation, reciprocal royalties) that relate directly to competition in the product market may not be sufficient to save it. As noted above, the antitrust agencies in the United States would find a significant diminution of rivalry in either the 'R&D market' or the 'technology market' equally unacceptable. In its challenge of the aircraft manufacturers patent pool, for example, the US Department of Justice argued that a royalty-free patent cross-licensing agreement among aircraft manufacturers which did not constrain competition with respect to prices, quantities, markets or channels of distribution nevertheless lessened competition substantially in the introduction of technological improvements in aircraft (Bittlingmeyer, 1988). While he has not yet stated his position with respect to the role that the concepts of R&D and technology markets might play in the enforcement of the Competition Act in Canada, the importance attached by the Director to the diminution of new product rivalry as a consequence of a merger between daily and community newspapers in Southam implies that enforcement policy in Canada will ultimately follow the US lead.[10]

Notwithstanding their apparent exclusion, static and dynamic efficiency considerations might be introduced indirectly into section 45 proceedings in support of an argument that, in the absence of the agreement, none of the parties involved would have entered the market and, as a consequence, the agreement cannot have lessened competition unduly. That is, absent the agreement, the product or sequence of products would not exist and, relative to that benchmark, the agreement does not lessen competition. Arguments to this effect are accepted by the courts in the United States.[11] This type of argument might also be used in cases where a sequence of innovations would be substantially delayed in the absence of an agreement. There does not appear to be any scope, however, for consideration of arguments that innovative rivalry is duplicative and/or precipitous and that an agreement reducing R&D spending and/or slowing a patent 'race' may be socially beneficial.[12]

Section 45 also appears poorly suited to the conduct of rule of reason analysis of individual restraints that are part of broader licensing or other agreements, as is possible in the United States. A restraint that lessens competition substantially would be difficult to defend on the grounds that it is reasonably necessary to align the incentives of the licensee and licensor. A possible defence might be that a restraint that is reasonably necessary to achieve an efficiency enhancing objective would not meet the objective intent requirement for undueness that is now part of the jurisprudence under section 45.[13]

The information shared in by the parties in a strategic alliance is likely to extend beyond formal R&D to production and even to marketing (in the form of customer experience and suggestions). The decision of the Supreme Court of Canada in *R.* v. *Nova Scotia Pharmaceutical Society* (*PANS*) raises the possibility that an agreement involving the exchange of information relating to customer experience and production experience, as well as to technology, might be sufficient to infer the existence of a tacit agreement to lessen competition which would contravene section 45 if the parties involved are large relative to the market.[14] The decision in *PANS* suggests a willingness to infer an undue lessening from agreements stopping short of naked restraints if the parties involved have substantial market power. A defence that sharing information about customer and production experience reduced the cost of innovation would not be entertained. If, however, the same evidence were presented in support of an argument that cooperation had the effect of increasing the rate of quality improvement and is pro-competitive, at least in part, this might go to the issue of undueness and thus be taken into account.

Collins and Hunter (1994) comment in this regard that, in general, market power is likely to be short-lived and thus inferences of undueness difficult to support in industries characterized by rapid product and process evolution. Baumol and Ordover (1992) also make the point that the current product market status of the parties in the agreement should be given less weight and the market for the technology given more weight when drawing inferences regarding the market power of a strategic alliance. Contrary to this, the *Antitrust Guidelines for the Licensing of Intellectual Property* appear to give equal weight to the product, technology and innovation markets (unless innovation and technology markets cannot be defined) and would find a lessening in *any one of them* to be cause for concern. A more explicitly dynamic approach might require a lessening of competition *in all three* of these markets (if they can be defined) in order to support an inference of undueness. If the technology and/or the innovation market cannot be defined, inferences regarding product market power should be supported by evidence on entry, exit, the variability of market shares and the existence of substitute products as well as traditional concentration evidence. This kind of evidence may be difficult to obtain, however, in situations where product evolution is rapid.

Strategic alliances are also reviewable as an abuse of joint dominance under section 79 of the Act.[15] Given that strategic alliances generally involve swaps of intangible assets, they necessarily exclude parties with nothing to swap and may therefore be regarded by the Director as foreclosing the supply of a scarce input (technology). In this, the Director might be urged on by potential imitators seeking to force established firms to licence core technologies. The superior performance and intellectual property defences under sections 79(4) and 79(5) respectively might be employed here.

CONCLUSIONS

The new approach to industrial policy assigns government the role of a facilitator of and sometime participant in collaborative arrangements for the purposes of innovation. As a facilitator, the government must ensure that the legal framework is such as to accommodate and encourage efficient institutional interaction.

It is generally believed that the enforcement of competition law in Canada poses few obstacles to interaction among innovative firms. A number of the sources of concern in the United States have involved private actions, and these have not been problems in Canada to date.[16] While the Director's policy statement, *Strategic Alliances Under the Competition Act,* together with some interpretations of the Supreme Court's decision in *R.* v. *Nova Scotia Pharmaceutical Society* may have shaken the view that competition policy will be accommodating to cooperative innovation, concerns are speculative at this point.

While the Sherman Act arguably offers more flexibility than the Competition Act in dealing with strategic alliances, the difference may be more apparent than real. Firms in strategic alliances with potential problems under section 45 may be able to make creative arguments regarding both the factors supporting an inference of undueness and the requirements for objective intent. Moreover, given the general requirement in US merger law that efficiencies be price-reducing as opposed to merely welfare-enhancing, the efficiencies defence for horizontal restraints may not apply frequently.

NOTES

1 Magun finds that 63 per cent of the alliances in which Canadian firms are involved are with foreign firms (1996: 22).

2 This change in policy was announced in the February 1990 budget. In the April 1993 budget, the Minister of Finance stated that the policy of requiring repayment of most grants and contributions to business by federal departments 'reinforces the government's efforts to orient its business assistance towards investing in economic development rather than the subsidization of the private sector'

3 Mowery and Rosenberg (1989: 33) write:

> The sequence of technological knowledge preceding scientific knowledge has by no means been eliminated in the twentieth century. Much of the work of the scientist today involves systematizing and restructuring in an internally consistent way the knowledge and practical solutions and methods previously developed by the technologist. Technology has shaped science in important ways because it acquired some bodies of knowledge first and, as a result, provided data that in turn became the 'explicanda' of scientists, who attempted to account for or to codify these observations at a deeper level.

4 Rosenberg (1982: 75) writes:

> Often, however, an innovation from outside will not merely reduce the price of the product in the receiving industry but will make possible wholly new or drastically improved products or processes. In such circumstance it becomes extremely difficult even to suggest reasonable measures of the payoff to the triggering innovation because such innovations open the door for entirely new economic opportunities and become the basis for extensive industrial expansion elsewhere.

5 Given the cumulative nature of the innovative process, it may not be appropriate to speak of an originating or pioneering innovator. This raises the interesting question of the stage at which an innovation passes into the domain of open science and becomes freely available.

6 The relationship between the magnitudes of $A+B$ and L is given by the equation:

$$\frac{L}{A+B}=\frac{1}{2}\left[\frac{c_0-c_2}{c_0}\right]e$$

where e is the price elasticity of demand. If, for example, $e=2$ and the sequence of innovations would have reduced unit cost by 20 per cent, then L is one-fifth of $A+B$.

7 For a historical survey of the treatment of intellectual property under competition legislation in Canada see Anderson *et al.* (1991).

8 The competitive impact of post-expiry royalties would depend on whether the patentee and licensees would be competing after the expiry of the patent.

9 In economic jargon, these agreements must satisfy the participation constraint. See Lewis and Yao (1995).

10 For a discussion of the importance attached by the Director of Investigation and Research to product innovation rivalry in the *Southam* case, see McFetridge (1996).

11 For example, in *Northrop* v. *McDonnell Douglas Corp.* the Ninth Circuit Court found that a 'teaming agreement' that allocated markets for a new aircraft was necessary for the development of the aircraft and was thus potentially pro-competitive rather than anti-competitive. See Holmes (1994: 556–57). Arquit and Kattan (1991) note that the creation of a new product has been accepted by the courts as a cognizable efficiency since *Broadcast Music Inc.* v. *Columbia Broadcasting System* in 1979. If it can be shown to have an efficiency rationale, a horizontal agreement is then evaluated on a rule of reason basis with the test being whether the agreement results in lower prices, better products or new products.

12 As noted earlier in the text, the Department of Justice/Federal Trade Commission *Antitrust Guidelines for the Licensing of Intellectual Property* also take it as given that more R&D spending and a faster rate of innovation are always better (section 3.2, Example 4).

13 To establish objective intent, it must be shown that a reasonable business person would or should have known that the agreement in question would likely prevent or lessen competition unduly. It may be possible to argue that when some of the objectives of an agreement are potentially efficiency enhancing, objective intent is not present.

14 *R.* v. *Nova Scotia Pharmaceutical Society*, [1992] 2 S.C.R. 606; 93 D.L.R. (4th) 36.

15 The abuse of dominance example given in *Strategic Alliances Under the Competition Act* (pp. 21–2) involves a traditional vertical squeeze in which an

upstream firm that is vertically integrated by means of a strategic alliance with a downstream firm denies supply of a patented input to a competing, non-integrated downstream firm. Since this squeeze reduces and may even eliminate the double mark-up problem that occurs when there is monopoly at one stage of production and duopoly at another, it might be socially beneficial in a static context. The case should also be assessed from the perspective of the incentive to invent.

16 Virtually all of the recommendations of Baumol and Ordover (1991) for making US antitrust law less hostile to innovation involved reduction in the scope for private actions.

BIBLIOGRAPHY

Anderson, R. D., Khosla, S. D. and Ronayne, M. F., 'The Competition Policy Treatment of Intellectual Property Rights in Canada: Retrospect and Prospect', in R. S. Khemani and W. T. Stanburry, eds, *Canadian Competition Law and Policy at the Centenary* (Halifax, Nova Scotia: Institute for Research on Public Policy, 1991).

Arquit, Kevin J. and Kattan, Joseph, 'Efficiency Considerations and Horizontal Restraints', *The Antitrust Bulletin*, Volume 36, Winter 1991.

Baumol, W. and Ordover, J., 'Antitrust: Source of Static and Dynamic Inefficiencies?', in T. Jorde and D. Teece, eds, *Antitrust, Innovation and Competitiveness* (New York: Oxford University Press, 1992).

Bittlingmeyer, G., 'Property Rights, Progress and the Aircraft Patent Agreement', *Journal of Law and Economics*, Volume 31, April 1988.

Collins, P. and Hunter, L., 'Strategic Alliances under the Competition Act: A Commentary on the Director's Draft Information Bulletin', *Canadian Competition Record*, Volume 15, Autumn 1994.

David, Paul A. 'Knowledge, Property and the System Dynamics of Technological Change', *Proceedings of the World Bank Annual Conference on Development Economics 1992* (Washington, DC: International Bank for Reconstruction and Development, 1993).

Director of Investigation and Research, Competition Bureau, *Strategic Alliance Under the Competition Act* (Ottawa: Industry Canada, 1995).

Gilbert, R. and Sunshine, S., 'Incorporating Dynamic Efficiency Concerns in Merger Analysis: The Use of Innovation Markets', *Antitrust Law Journal*, Volume 63, Winter 1995.

Grossman, G., 'Promoting New Industrial Activities: A Survey of Recent Arguments and Evidence', *OECD Economic Studies* (Paris: OECD, 1992).

Harris, R., 'The New Protectionism Revisited', *Canadian Journal of Economics*, Volume 22, 1989.

Holmes, W., *Antitrust Law Handbook* (1994 edn) (Deerfield: Clark, Boardman, Callaghan, 1994).

Katz, Michael L. and Ordover, Janusz A., 'R&D Cooperation and Competition', *Brookings Papers: Microeconomics* (137–203) (Washington, DC: Brookings Institute, 1990).

Kennish, T., 'Competition Act Treatment of Strategic Alliances', *Osler, Hoskin and Harcourt Competition Law Review* (Toronto: Osler, Hoskin and Harcourt, Summer 1994).

Khemani, R. S. and Waverman, L., 'Strategic Alliances: Implications for Competition Policy', Paper presented to the Competition Law and Policy Committee of the OECD, November 1992, Paris.

Krugman, P., 'Competitiveness: A Dangerous Obsession', *Foreign Affairs*, Volume 73, March/April 1994.

Kumar, V. and Magun, S., 'The Role of R&D Consortia in Technology Development', *Occasional Paper Number 3* (Ottawa: Industry Canada, 1995).

Lewis, T. and Yao, D., 'Some Reflections on the Antitrust Treatment of Intellectual Property', *Antitrust Law Journal*, Volume 63, Winter 1995.

McFetridge, D. G., 'The Economics of Mergers and Abuse of Dominant Position', Paper Prepared for the Canadian Bar Association, National Competition Law Section, 1996 Annual Competition Law Conference, Ottawa, 27 September.

Magun, S., 'The Development of Strategic Alliances in Canadian Industries: A Micro Analysis', *Working Paper Number 13* (Ottawa: Industry Canada, 1996).

Merges, Robert P. and Nelson, Richard R., 'On the Complex Economics of Patent Scope', *Columbia Law Review*, Volume 90, 1990.

Miller, E., 'Antitrust Restrictions on Trade Secret Licensing: A Legal Review and Economic Analysis', *Law and Contemporary Problems*, Volume 52, 1989.

Mowery, D. and Rosenberg, N., *Technology and the Pursuit of Economic Growth* (New York: Cambridge University Press, 1989).

Nelson, R., *National Innovation Systems* (Oxford: Oxford University Press, 1992).

Niosi, J. and Manseau, A., 'Canada's National R&D System' (Mimeo) (CIRST, Université du Québec à Montréal, 1994).

Ordover, Janusz A., 'A Patent System for Both Diffusion and Exclusion', *Economic Perspectives*, Volume 5, Number 1, 1991.

Ordover, Janusz A. and Willig, Robert D., 'Antitrust for High-Technology Industries: Assessing Research Joint Ventures and Mergers', *Journal of Law & Economics*, Volume 28, 1985

Pitofsky, R., 'Proposals for Revised United States Merger Enforcement Guidelines in a Global Economy', *The Georgetown Law Journal*, Volume 81, December 1992.

Porter, M., *The Competitive Advantage of Nations* (New York: The Free Press, 1990).

Porter, M., *Canada at the Crossroads* (Ottawa: Supply and Services, 1991).

Rosenberg, N., *Inside the Black Box: Technology and Economics* (New York: Cambridge University Press, 1982).

Scotchmer, Suzanne, 'Standing on the Shoulders of Giants', *Economic Perspectives*, Volume 5, Number 1, 1991.

Trebilcock, M., Chandler, M. and Howse, R., *Trade and Transitions* (New York: Routledge, 1990).

United States Department of Justice and the Federal Trade Commission, *Antitrust Guidelines for the Licensing of Intellectual Property* (Washington, DC, 1995).

Williamson, O., 'Credible Commitments: Using Hostages to Support Exchange', *American Economic Review*, Volume 83, September 1983.

10

COMPETITION AS COOPERATION

SEMATECH, Inc. and the case for competitive industrial policy

Richard J. Braudo

INTRODUCTION

The Age of Global Competition is upon us, not only between transnational corporations (TNCs), but also between regional trading blocs, and between and among their member nations. In the Information Age of the late 1990s and twenty-first century (global markets, regional trading blocs and TNCs in place), it is difficult to imagine that there are not significant implications for existing domestic and international competition law regimes, including especially patent law regimes.

The emergence of TNCs as pre-eminent players in international trade can be traced back to US-led reconstruction following the Second World War. In Western Europe and Japan, cooperative industrial policy was orchestrated through sectors supported by US-based governmental and corporate operations; i.e. targeted, US-based international cooperative industrial policy. In both war-torn capitalist Europe and Japan, massive American public and private capital investment, combined with manufacturing, marketing and organizational know-how, provided the foundation for non-military reindustrialization. As a result, perhaps not surprisingly, the Allies' principal enemies, Germany and Japan, emerged to become leading global economic rivals by the 1970s.

Arguably, the rise of German and Japanese competitiveness (if not competitive advantage) is largely attributable to the fact that both economies had no choice but to proceed after the Second World War to focus domestic industrial development exclusively in the non-defence arena (whereas, it is only since the late 1980s and end of the Cold War, that Canadian, US and other Western European economies have become predominantly focused on non-defence development).[1] The current downsizing of defence sectors falls short of the virtual dominance of the non-defence sectors in

Germany and Japan since 1945. In addition to the non-military orientation and US-based cooperative participation in their overall industrial efforts, competitive edges in Germany and Japan may have resulted from industrial policies distinct from those typically practised elsewhere. In particular, in the context of a maturing global economy, rather than being anti-competitive, one theme of this chapter is that cooperation between government and corporations, or among corporations themselves, can be seen to foster efficiency and competitiveness.

A second, and related theme, is that economic growth in an era of globalization has been, and will increasingly be, driven by technology and innovation. There is general consensus that the revolution in information processing, including the merging of telecommunications and data processing technologies through creation of so-called 'super-highways', is the crucial factor leading to the emergence of the global marketplace in the twenty-first century:

> If the revolution in transportation and communications of the mid-nineteenth century created modern global competition, the second communications [information processing] revolution of the mid-twentieth century has probably been the major force in shaping its current configuration.
>
> (Chandler, 1986: 442)

Accordingly, the focus of this chapter is on the role played by high technology-driven TNCs[2] in the global marketplace, TNC interaction with government, and the resulting implications for prevailing competition policy and intellectual property (patent law) regimes:

> improving the competitiveness of the United States [and Canada] will involve far more than increasing the profitability of [North] American companies or attracting more foreign investment. It will require a radical reevaluation of the fundamental assumptions and institutions of [North] American capitalism. But before this process can proceed, we must first ask ourselves . . . what it means to be 'competitive' and how much we are willing to sacrifice to achieve this goal.
>
> (Blaine, 1996: 116–17)

> Structural competitiveness is decidedly not simply 'getting prices right', but a much more complex mix of getting prices right, matching societies and institutions smoothly, and building national and regional networks of interdependent economic interactions.
>
> (Crone, 1996: 2)

A review of empirical evidence (derived from a survey of relevant literature) and a case study of SEMATECH, Inc. (Austin, Texas; based on in-person interviews with management personnel and various SEMATECH-specific publications) provide the groundwork for the original analysis leading to the key theme of the chapter:

> Prevailing economic, legal and political conditions call for re-evaluation of the conceptualization of competition which currently underlies the development and implementation of competition policy in Canada and the US, if not world-wide.

The chapter concludes that the post-Cold War reinvention of government is a necessary but by no means sufficient first step towards establishing or maintaining global competitiveness. Just as necessary, perhaps requiring a quantum leap in law and economics thinking underlying public policy-making for the twenty-first century, is the need to reconceptualize competition ('New Competition'), incorporating cooperation as a key element of the New Competition:[3]

> cooperation occurs when people or groups act together in a co-ordinated way to pursue shared goals, enjoy an activity, or simply further their relationship . . . Cooperation usually involves not only the coordination of activities but also the sharing of the benefits that emerge from the cooperation . . . Cooperation is also said to exist when behavior 'maximizes both the individual's and others' interests, whether the situation involves correspondent or noncorrespondent interests' . . . Individuals engaged in truly co-operative behavior tolerate conflict and use real-time communication . . . and individual give-and-take discussions to reach productive outcomes . . . Finally, a cooperative relationship casts a 'shadow of the future': parties treat each other as though their future relationship counted . . .
>
> (Browning *et al.*, 1995: 123)

TNCs: THE PRINCIPAL PLAYERS IN GLOBALIZATION

While marginal growth in employment in North America during the past several years has been attributed to activities of small and medium-sized firms (SMEs), where global competition is concerned the dominant actors continue, increasingly in fact, to be TNCs:

> An understanding of the evolution of modern global competition requires, therefore, a description and analysis of the evolution of this new species of business enterprise, the new player in international trade – the integrated, diversified multinational [TNCs].
> (Chandler, 1986: 406)

Moreover, it is high-technology TNCs which are the primary players on the field of global competition:

> The multinationals of today are not searching for resource inputs as in the earlier portion of the twentieth century . . . but are spreading the economies of scale and scope of increasingly technology driven production across borders and among firms. *This increasing globalization of product and capital markets, combined with the accelerating rhythm of technological innovation has significantly altered the competitive environment of the large corporate enterprise [TNCs].*
> (Comanor *et al.*, 1992: 3; emphasis added)

It is beyond the scope of this chapter to delve into the evolution of TNC pre-eminence on the global stage. Nevertheless, it is important to consider the instrumental role of high-technology TNCs to emerged and emerging global markets. One significant example of this development is the composition of joint ventures since the 1980s:

> There has evidently been a decade-long trend in the distribution of joint venture formations away from their traditional focus in the energy, chemicals and metal industries, towards the computer, electronic components, communications systems, pharmaceuticals, medical equipment, and financial services industries.
> (Shapiro and Willig, 1990: 117)

> the recent upsurge of collaborative alliances represents a qualitatively new stage, with a much greater *number* of agreements, an increased *range* of industrial sectors involved, and far more *international* collaborations. In particular, the collaboration in the *production of technological knowledge*, with the partners' dependent on each other's complementary skills and assets, is a relatively new phenomenon . . .
> (Coombs *et al.*, 1996: 6)

Another illustration of the importance of high-technology TNCs in global competition is the increased incidence of mergers between them, both within and across borders.[4]

In other words, industrial reorganization and growth is driven primarily by high-technology and directly related industries. Consistent with the focus on high-technology TNCs is general consensus that, in a global marketplace, technological innovation and its commercialization are the most important ingredients for economic growth:

> in the existing world, the role of technological innovation and its diffusion among nations is crucial to the development of the international economy. Equally, TNCs, concentrated as they are in technology-intensive activities, constitute the major medium through which technology is created and the major vehicle of its international diffusion.
>
> (Gray, 1992: 173)

The dominance of TNCs in respect of globalization is further evidenced by levels of foreign direct investment (FDI), the incidence of strategic alliances, and trends in international trade. Since 1980, global FDI has been growing faster than international trade and strategic alliances have been occurring in the tens of thousands. In both FDI and strategic alliance formation, the leading participants are TNCs.

Moreover, an increasing proportion of global trade is the result of intermediate goods and services produced and traded within TNCs:[5]

> it can be anticipated that transnational networks and transborder alliances, already a major factor in international economic flows, will grow in importance.
>
> (Vernon, 1992: 26)

> a large and growing proportion of world trade involves intra-enterprise trade. For Japan and the United States, for example, trade related to FDI now accounts for over half of the total trade flows.
>
> (Ostry, 1992: 9)

> the fact is that between 30 and 40 per cent of the sales of the leading industrial TNCs are now produced outside of their national boundaries, and that the value of those sales now considerably exceeds that of international trade.
>
> (Dunning, February 1992: 142)

The tentacles of TNCs reach niches throughout the global marketplace, including those in which SMEs participate. The dominant role of TNCs in global trade, high-technology industrial reorganization, innovation and growth is matched by their primary role in terms of economic output itself:

> adding the actual value-added of foreign-based TNCs in a country to that of the foreign output of its home-based TNCs and expressing the result as a percentage of private GNP (GNP minus government expenditures), the resulting percentages currently exceed 50 per cent in the cases of such economies as Belgium, Canada, Hong Kong, Netherlands, Singapore, Switzerland and the United Kingdom; more than 30 per cent in the cases of Australia, France, Germany and Italy; and more than 20 per cent in the cases of Brazil, Japan, Nigeria, Taiwan Province of China and the United States . . . these rates have been rising over the past ten years and seem likely to continue to rise in the next decade.
>
> (Dunning, December 1992: 12)

It is not an exaggeration to conclude that high-technology TNCs play a direct role or exert major influence in virtually every aspect of economic life globally.

The universal impact of TNCs is evident not only from their primary role in trade, industrial reorganization, innovation, growth (as well as leading participation in capital markets), but also from their significant influence in respect of the development and movement of human capital:

> Information, brainpower and organizational capacity (all primary components of TNCs) are excellent examples of created intangible assets; but as human beings become better qualified and more affluent, and as real travel costs fall, the cross-border movement of people is also likely to increase . . . TNCs are one of the main vehicles making this asset more mobile, particularly within their own organizations and between advanced industrialized countries.
>
> (Dunning, December 1992: 14)

In addition, contrary to public perception, TNCs are based not only in North America, Europe and Japan (the Triad), but throughout the industrialized world:

> every industrialized country provides a base for a considerable number of TNCs . . . even the larger firms in some of the rapidly industrializing countries of Asia and Latin America had joined the trend.
>
> (Vernon, 1992: 7)

This brief survey of the nature and role of TNCs (high-technology TNCs in particular) attests to their pre-eminent importance in the global economy. Therefore, the behaviour of TNCs in respect of their private market activities is of primary concern to any nation faced with issues of global competitiveness, as are the implications for public policy of the development of TNC relationships with government.

TNCs AND GOVERNMENT: EMERGING PARTNERS IN ECONOMIC GROWTH

The history of corporate and government relations is long and varied, beginning with state-chartered companies in the sixteenth and seventeenth centuries. The *raison d'être* of such companies was the creation of economic infrastructure (primarily transportation) deemed vital to mercantilist growth. This (preclassical) relationship was like that of king and prince. Corporations were granted the exclusive right to create value for themselves, the royal family and extended family members (royal subjects) within the strict constraints of their state-defined corporate charters.

By the nineteenth century, however, the classical limited liability company emerged as the dominant corporate structure. Close-knit common bonds of state-sanctioned monopoly had been replaced by the classical view that monopoly grants were undesirable. Instead of official monopoly rights and government control through narrow corporate charters, corporations were free to expand as the market permitted, controlled from within by their directors. In the absence of government regulation/competition policy (classical *laissez-faire* economics being the order of the day), corporations grew rapidly.

Towards the end of the nineteenth century, conglomerates with significant domestic monopoly power had emerged. As a result, a new relationship between government and corporations developed, one characterized by regulation, if not hostility, and embodied, for example, in passage of the Sherman Anti-Trust and Clayton Acts by the US Congress in 1914. Notwithstanding unfriendly Canada/US competition/anti-trust regimes, neoclassical corporations grew to become modern multinational corporations, many of which have survived to become TNCs in the 1990s. It is fair to say that in Canada and the US, government attitudes toward corporations, embodied in prevailing competition and anti-trust[6] regimes, originally formulated to undo and prevent monopolistic power in a non-global era,[7] have remained basically unchanged. TNCs continue to be monitored closely and regulated, with civil and criminal enforcement, if need be, the order of the day.

Relations of TNCs with governments in many other countries have been quite different. In this regard, it is important to keep in mind that while TNCs might be headquartered in North America, their relations with governments in other nations are not dictated by their home government's attitudes or policies towards them. At first, when TNCs entered countries in need of the economic development offered, TNCs were positioned to dictate the terms of the relationship. By the 1970s, however, host countries consisted of developed as well as developing countries. During the last decade, at least at the provincial and state level in Canada and the United States, and throughout most of the industrialized world, the relationship is better characterized as being one negotiated based on mutually beneficial interests.[8]

The evolution of TNC–government relationships has taken place together with the advent and growth of globalization. As shown in the discussion above, TNCs increasingly play leading roles on the global economic stage. Regulation and control, adjudication and prosecution can hardly be described as being consistent with mutually beneficial relations based on negotiation:

> the influence that individual Governments are in a position to exert over their respective transnational corporations appears rapidly on the decline. *Although Governments have been known to remain blind to the obvious for remarkably prolonged periods of time*, that one eluctable fact should eventually lead them to limit their unilateral efforts at control.
>
> (Vernon, 1992: 27; emphasis added)

Nevertheless, in Canada and the US, federal governments continue to behave according to the hostile attitude dictated by a bygone era.

Whether the Canadian and US governments like it or not, the dominant position, economic strength, and allocative flexibility of TNCs necessitate a change towards mutually beneficial relations at all levels:

> The highly concentrated nature of many global industries, in some cases populated by but a handful of global competitors, has also shifted the context of the interface between governments and business from regulation to negotiation.
>
> (Doz, 1986: 226)

This does not mean that TNCs should be permitted or are in a position to dictate relations with government. On the contrary, even in concentrated industries, Canadian- and US-based TNCs face serious competition from global rivals. It is because of such competition that the need has arisen for changed relations of Canadian and US governments with TNCs, based on mutually beneficial negotiation.

Government influence on TNCs and people residing in its jurisdiction is no less significant than the effect of TNCs on government and people living in their locales. For the Canadian and US governments to continue operating as if neoclassical markets, preserved with the application of prevailing competition regimes, are necessary (and even sufficient) to maintain or improve national competitiveness globally is 'to remain blind to the obvious'.[9] Given the increasing importance of TNCs, relationships between TNCs and governments outside of North America, and global challenges to Canadian and US competitiveness, the need to transform government and TNC relations in North America is striking.

To replace the neoclassical posture of a bygone era and the *laissez-faire* approach championed since the 1980s, the forging of a new partnership of cooperation is called for, one through which common economic goals based on global competitiveness are achievable:

> For better or for worse, throughout the world, national Governments are fashioning the organization and structure of value-added activities within and beyond their borders in a host of ways. *Any Government of the 1990s that ignores this fact and pursues a 'hands-off' or 'leave-it-to-the-market' strategy is likely to be as negligent in promoting the welfare of its citizens, as were its predecessors of the 1960s and 1970s, that sought to replace the discipline of the market by socialist or centrally planned macro-organizational policies.*
>
> (Dunning, December 1992: 8; emphasis added)

Such partnership, moreover, must not ignore the approach to government–TNC relations taken by global rivals in a variety of high-technology industries, including semiconductors:

> Governments in Japan, Europe, and South Korea have supported their semiconductor industries through technology policies and/or financial support. For example, Europe's JESSI consortium plans to spend about $14 billion over the next 5 years on semiconductor R&D.
>
> (GAO, 1992: 27)

Fortunately for the United States (and hopefully Canada), SEMATECH, Inc., is a well-established, remarkable experiment in the creation of such a partnership. The discussion now turns to an evaluation of SEMATECH's performance to date.

SEMATECH, INC.: CASE STUDY OF US COOPERATIVE INDUSTRIAL POLICY

If necessity is the mother of invention, SEMATECH, Inc. was duly invented. SEMATECH (SEmiconductor MAnufacturing TECHnology) is 'a government–industry R&D consortium formed in 1987 (with enabling

U.S. Congressional legislation under the National Cooperative Research Act of 1984) to enable the United States to regain world leadership in semiconductor manufacturing' (GAO, 1992: 1). Its mission 'is to provide the U.S. semiconductor industry with the domestic capability for world leadership in manufacturing' (GAO, 1992: 8).[10] It is a cooperative, non-profit joint venture consortium which initially included Advanced Micro Devices, Inc., AT&T, Digital Equipment Corporation, Hewlett-Packard Company, Intel Corporation, IBM, Motorola, Inc., National Semiconductor Corporation, Rockwell International Corporation, Texas Instruments, Inc., and the US Defense Advance Research Projects Agency (DARPA). Funding is provided on an annual basis, with total member company contributions matched by DARPA, up to $100 million annually:[11]

> the founders of SEMATECH came to recognize that a new order was required in their industry to avoid its demise. Their solution showed how open they were to radical change. They decided to do what they had never done before – to found a new organization form, previously unavailable to them because of legal restrictions, that would facilitate *cooperation* [emphasis added] in solving common problems. They backed that decision up with substantial commitments of their own financial resources.
>
> (Browning *et al.*, 1995: 142)

Member contributions are $15 million maximum (IBM, Intel, Motorola and Texas Instruments) and $1 million minimum, with the financial contribution formula based on member companies' semiconductor sales revenues. Assignments of member company personnel to work at SEMATECH are based on the percentage of total contributions made by a given firm, and must be at least 1 per cent and not more than 15 per cent of total SEMATECH assignees. Thus, SEMATECH was created as a full-fledged partnership in terms of both monetary (liquid) and human capital provided by its principal founders (TNCs and government).

SEMATECH functions as a true consortium, independent of control by any one member, including DARPA:

> One reason that SEMATECH has been able to operate as a true consortium is the government has not assumed the role of an overall director or manager.
>
> (SEMATECH, 1990: 5)

Moreover, SEMATECH-related cooperation is not limited to member participation. The land on which it is located is owned by the University of Texas and leased to SEMATECH for $1/year. SEMATECH has also established a national network of research centres involving thirty-nine universities and research institutions (SEMATECH Centres of Excellence, SCOE). SEMATECH provides these SCOEs annual grants for long term research totalling about US$10 million. In addition to augmenting SEMATECH's basic research efforts, the SCOE program facilitates an increase in the number of engineers with graduate degrees and experience related to the semiconductor industry.[12] Finally, Semi/SEMATECH, Inc. was created to serve as a formal communications link between US semiconductor industry suppliers and SEMATECH and its member companies:

> According to its participants, Sematech has fostered their gains in part by improving communication between the American semiconductor industry and the American semiconductor equipment industry, which has traditionally been fragmented, disputatious, and undercapitalized, especially in comparison with its Japanese counterpart.
>
> (Tyson, 1993: 153)

In short, SEMATECH endeavours to harness cooperation in all available ways necessary to the accomplishment of its mission.[13]

Given that its mission is competition-driven, based on R&D and innovation, critical to its success is SEMATECH's development and treatment of new technology. A full understanding of these aspects of its efforts requires awareness of two basic principles underlying SEMATECH's creation. First, contrary to what some observers may perceive, its *raison d'être* was more industry- than government-driven. From the outset, member companies have provided at least half of the operating funds, the core management and technical personnel.[14] SEMATECH is run accordingly, with management having demonstrated a willingness and ability to make changes called for by market conditions. Second, SEMATECH was created to bring together and champion three levels of organization critical to the innovation vital to achieving its mission. Through its unique ability to combine expertise in (1) strategic thinking, (2) project proposal development and (3) project implementation, based on an industry-wide, global perspective, SEMATECH has proved to be successful at all three levels of organizational activity.

In addition to the two underlying principles identified above, SEMATECH's development of a quality management program ('Total Quality Initiative'), and its program for treatment of patent rights associated with innovation of technology, has resulted in its assumption of a leadership position in the US Semiconductor Industry Association. In any innovative endeavour linked to manufacturing, particularly semiconductor fabrication, quality control is a primary ingredient to success in the marketplace. Through the cooperation of its members and 'extended family'[15] depicted above, SEMATECH has internalized the ingredients needed for effective innovation.[16]

SEMATECH's effectiveness also depends on the ability to confer benefits beyond its members and direct suppliers. In this respect as well, it has proved worthy of good marks:

> U.S. semiconductor manufacturers that are not members of SEMATECH can benefit from much of SEMATECH's R&D program because its primary objective has been to strengthen the U.S. semiconductor equipment supplier base . . . In addition, SEMATECH is the focal point for (1) improving long-term relationships between semiconductor manufacturers and their key equipment and material suppliers through its Partnering for Total Quality program and (2) developing industrywide standards for semiconductor manufacturing equipment through its total cost ownership model, equipment qualification and characterization procedure, CIM [computer integrated manufacturing], and flexible manufacturing programs.
>
> (GAO, 1992: 16)

Such cooperative innovation has been a crucial component of SEMATECH's contribution to the resurgence in the 1990s of US competitive advantage in semiconductor manufacturing.

Given SEMATECH's success to date, to argue that its approach has discouraged innovation[17] would seem to be an ideological claim lacking empirical foundation. At the microeconomic level, for fiscal year 1992, ten member companies received more than a 200 per cent return on investment ('ROI' for capital contributed), with one gaining only a 150 per cent return.[18] On the macroeconomic level, by the end of 1993, in terms of world-wide market share, the US semiconductor manufacturing industry had re-established its position as world leader. The evidence shows that in terms of total investment, rate of investment, ROI and overall industry performance, participation in and cooperation with SEMATECH has significantly surpassed results obtained from members of the industry competing without such coordination. Through its contributions to improved manufacturing technology, among other things, SEMATECH has played an instrumental role in the US recovery of global semiconductor market leadership.[19]

A review of its performance confirms that 'SEMATECH is demonstrating that despite the varied interests of its members, government and industry can work together with a high level of consensus in an atmosphere of pre-competitive cooperation' (SEMATECH, 1990: 10). So far, the experiment has proven itself to be more than just a factor in the winning or losing of global market share in the semiconductor industry.[20] As a case study, SEMATECH illustrates both the need for and a workable approach to a reconceptualization of competition ('New Competition'). One purpose of the New Competition is to develop competition policy which serves the interests of TNCs, government and related parties faced with the challenges of global competition in the mature Age of Information of the twenty-first century.[21]

THE SEMATECH EXPERIENCE: IMPLICATIONS FOR PATENT LAW REGIMES

This section explores possible implications of SEMATECH for the underlying rationale and structure of patent law regimes prevailing in major industrialized countries like Canada and the US. More specifically, questions are raised relating to the tension in patent law between competing private and social needs, inefficiencies associated with patent races, and incentive and efficiency problems arising from the cumulative nature of technology-based innovation.

Initially, SEMATECH, Inc. retained ownership of patentable technologies (process patents) or products (manufacturing equipment) developed, while member firms held the exclusive right to take new processes and products to market. Thus, commercial benefits attached to patent rights accrued exclusively to member firms. Coinciding with SEMATECH's success in developing new technology and contributing towards the recapture of global market leadership, has been the development of an understanding between management of SEMATECH and its member companies that achievement of SEMATECH's industry-wide mission depends on effective technology transfer. Two keys to success go hand in hand: effective innovation and marketing such innovations globally in the most effective manner possible. As a result, during the latter half of 1993, SEMATECH adopted a technology transfer strategy. The strategy allows for participation by non-member firms where such firms are determined to be the most effective party to market a given technology overseas.

Under the new technology transfer system, where a non-member subcontractor (typically, an emerging growth technology SME) develops a transferable technology (process or product), it is granted the patent and marketing rights. Meanwhile, member firms receive royalty-free licences to utilize the process or equipment.[22] This allows SMEs to market the innovative process or product outside the United States in competition with Japanese, Asian and European suppliers. As a result, firms too small or too specialized to participate inside SEMATECH are afforded the opportunity to benefit from their contributions – both by being paid for the cost of technology development and having access to potential profits from marketing their innovation(s) globally. At the same time, member firms gain access to new technology at no additional cost by not having to pay royalties and from knowledge of the innovations gained upon the return of their assignees.[23] Membership has its privileges as firms are afforded competitive advantages over their global rivals with respect to time, money and knowledge utilized in technology development and manufacturing.

In order to consider patent law regime questions raised by globalization in the technology sector, in light of SEMATECH, a summary of three issues frequently addressed in the literature on the law and economics of patents is appropriate:

1 Tension between competing private and social needs for innovation.
2 Inefficiencies associated with patent races.
3 Incentive/efficiency problems due to the cumulative nature of technology-based innovation.

The tension in patent law between competing private and social needs has received considerable attention in the literature on copyright as well as patent law:

> These [the patent] laws provide a right of exclusion as an economic incentive to the publication of creative work [process and product patents] . . . and perhaps because of a commonly-shared view that fairness requires some degree of protection. However important their objectives, they create legal rights which bar others from areas of competition and thus may conflict with anti-trust policy.
>
> (Scherer, 1983: 1)

As conventionally conceived, patent law regimes must balance the need to provide adequate private incentive to publicize innovation via patent issuance and the attendant rights of exclusivity against the need to promote competition via anti-trust or competition policy, inherently counter to the granting of exclusive legal rights to the patent holder.

The underlying dynamic which leads to the tension of differing private and public needs is the classic free-rider problem associated with public goods:

> Innovation and new technology have attributes of a public good in that use by one person does not preclude use by others and exclusion of free riders is difficult. This creates the twin social needs that the producer of new innovation be appropriately rewarded and that the new information be widely disseminated.
>
> (Brodley, 1990: 99)

According to neoclassical and modern economic analysis, private and social needs (incentives) are conceived as being distinct. Therein lies the public good attribution and free-rider problem, inherent tension and perception of innovation and new technology as requiring a broad-based and long-standing patent law regime. In the pre-Information Age, it is easy to understand why such a conception was appropriate.

It seems implausible, however, that in today's industrialized nations public dissemination of information/knowledge and technical know-how remains problematic. SEMATECH's approach to intellectual property, moving away from exclusive patent and towards 'common' market rights, indicates that the patent law component of competition policy needs to be re-evaluated. It suggests that the global competitive imperative, at least where cooperative R&D and technology transfer are concerned, renders restrictions on licensing on the one hand, and patent length on the other, less appropriate. This is surely the case where product cycles are short and first-mover advantages great, as is typically the case in high-technology industries. In an era of global competition, where lines are being drawn along national, international, interregional and global grounds, it is plausible to argue that the patent law concerns of yesterday have become barriers to competitiveness today and into the twenty-first century.

Perhaps one lesson to be learned from SEMATECH is that, in the maturing Age of Information and Global Competition, there is a need to develop post-modern law and economic analysis which incorporates new political, legal and economic realities. Through cooperative R&D, product development and product marketing ventures, which previously would have been considered anti-competitive and in violation of anti-trust law, innovation can be enhanced and national or regional competitiveness *vis-à-vis* the rest of the world improved. As SEMATECH's treatment of patents demonstrates, such effectiveness is achieved by moving away from exclusive rights and towards 'common' market rights. Given the microeconomic and macroeconomic results already identified, there is little basis to argue that the SEMATECH approach discourages innovation. The evidence suggests that in terms of total investment, rate of investment and ROI, SEMATECH has facilitated economic performance beyond the capability of members of the US semiconductor industry merely competing with each other.

This is not to say that because of competitive discipline provided by global competition, cooperative ventures ought now be seen as competitive *per se*:

> Antitrust authorities must be on the alert to detect ventures that offer the appearance but not the reality of independent action. For example, consider a consortium that produces a key input, say semiconductors, and sells this input to its members who independently set final output prices, say for personal computers.

> Suppose that the consortium encompasses the bulk of semiconductor manufacturing capacity and that entry into semiconductor production by non-members is very difficult. If, by joining the consortium, the members forsake the capacity to independently manufacture semiconductors, the consortium can simply raise the price it charges its members for semiconductors to a monopoly level. The fact that members compete in selling personal computers will not then prevent prices for personal computers from rising. The members will take their monopoly profits through their ownership of the consortium.
>
> (Shapiro and Willig, 1990: 11)

As Shapiro and Willig aptly illustrate, the potential for monopolistic control remains. In their example, however, given the presence of competitive foreign suppliers, assuming access to each other's markets and consortium-membership restricted to American firms, such monopoly action is unlikely. To the extent that the potential for monopolistic behaviour presents itself, rather than a *per se* prohibition of such cooperation, grounds arise for a rule of reason approach to the anti-competitive impact of consortiums. Given the SEMATECH experience, at least for domestic consortia, the more reasonable presumption is that of competitiveness rather than anti-competitiveness.

The effectiveness of SEMATECH in respect of quality control, rates of innovation, and ROI in R&D, suggests that resolution of the issue of inefficiencies associated with patent races is a key to improved global competitiveness. Historically, economists have viewed patent races, like competition in the marketplace, as efficiency enhancing. Analysis in recent years suggests, however, that patent races in many cases may have the opposite effect:[24]

> Baumol (1990) finds that these technology cartels have advantages over single competitors, seem likely to contribute to economic efficiency and that 'this may be one way in which anti-trust activity, as currently conducted (in the USA) does indeed inhibit innovation and productivity growth'.
>
> (Comanor *et al.*, 1992: 13–14)

Nevertheless, the perceived need for patent races continues to persist:

> Easing of antitrust rules reduces the constraints against cartels and monopolies. Moreover, absence of innovation rivalry prevents any standard for assessing relative innovation performance and thus for monitoring managerial behavior.
>
> (Brodley, 1990: 99)

Viewing SEMATECH and similar domestic or regional cooperative ventures[25] which are part of a globally competitive industry or marketplace as 'technology cartels' is substantively inaccurate. Cooperation and cartelization are not one and the same. They are distinct responses to a basic problem endemic to patent races:

> In reality, nothing guarantees that investment programs are made known to all concerned at the time of their inception. This uncertainty is especially high for the development and commercialization of new technology. Accordingly, innovating firms need to achieve greater coordination than the price system alone appears to bring about.
>
> (Jorde and Teece, 1990: 84)

Cartelization is an attempt to resolve duplication inefficiency via monopolization; i.e. by using coordinated action to manipulate the price system itself. SEMATECH-like cooperation, on the other hand, emphasizes trust-building as a prerequisite to resolving the information transfer problem. This dilemma, rarely addressed in the literature, appears to have been eliminated for the most part at SEMATECH. Resolution of the information transfer problem is seen by SEMATECH management as necessary to its achievement of so effective a rate of innovation.

The history of SEMATECH and the US semiconductor industry since 1987 demonstrates that, in a global marketplace with competitors from other regions of the world, cooperation does not mean cartelization. Instead, it can allow for reduction of inefficiencies due to duplication, barriers to information transfer and lost scale opportunities, all associated with patent race inefficiencies.[26] In the case of SEMATECH, created as a non-profit private sector–government cooperative venture, the benefits of taking advantage of economies of scale, sharing technological know-how and elimination of patent race inefficiencies are attainable. Moreover, such benefits accrue not by transferring social surplus from consumers to firms, but by creating real efficiency gains; i.e. creating increased private and social wealth. In addition, SEMATECH appears to have facilitated the sharing of increased corporate wealth between its TNC members and numerous SMEs.

In the context of global market conditions, the SEMATECH experience raises important questions, and may even provide answers to problems of tension due to competing public and private needs and inefficiencies due to patent races. Is SEMATECH instructive with respect to the issue of incentive and efficiency problems arising from the cumulative nature of technology-based innovation? The basic law and economics problem related to the cumulative nature of technology-based innovation is the determination of efficient patent breadth (scope) and length (years of exclusivity).

With respect to scope, incentive problems arise if too narrow a patent is granted, particularly in the case of new technologies. For example, it is often the case that early-stage innovation, perhaps the most difficult and costly to achieve, does not lend itself immediately to marketable technologies or products. Too narrow a patent regime in this context might discourage innovation that would ultimately have great social value. Alternatively, too broad a patent will encourage innovators to be first; i.e. exacerbate patent race problems. Moreover, such a regime will discourage further innovation by anyone but the primary inventor, which is especially problematic given the cumulative nature of technology-driven innovation.

In practice, the issue of patent breadth is litigated excessively.[27] SEMATECH suggests that in technology-driven, global markets, the willingness of member companies to forgo exclusive patent rights in exchange for the opportunity to be the first-to-market is highly significant. It shows that a competitive edge (the relatively narrow 6–9 month first-mover window of opportunity) for manufacturing, marketing and distribution of products derived from technology-driven R&D is sufficient incentive for (TNC) participation in such innovation. With respect to non-member firms, typically emerging growth companies (SMEs), the opportunity to participate in such innovation without benefit of participation in their home market but with the right to market the innovation overseas (the opportunity to open up new distribution channels) is sufficient incentive to participate.

It is evident, therefore, that patent breadth problems, including those of incentive and inefficiencies such as large costs stemming from litigation, are avoidable through cooperative R&D and product marketing ventures structured along the lines of SEMATECH. The dynamic relationship between patent breadth and cooperation, however, must also be considered:

> it is misleading to ask how broad patent protection should be without simultaneously asking whether research firms can integrate or otherwise cooperate. Similarly, it is misleading to ask how leniently society should treat cooperation among research firms without simultaneously asking how broad patent protection should be.
>
> (Scotchmer, 1991: 37)

As in the case of the tension between the need for public knowledge and exclusivity versus the desire for competition, problems arising from the breadth of patent issue may be avoided when innovation is conducted through a SEMATECH-like cooperative venture. This is not to say that such cooperative ventures warrant legal status *per se*. Rather, as with the tension between private versus public needs, it calls for a rule of reason approach in which competitiveness is presumed in favour of national or regional cooperative R&D ventures, with patent breadth seemingly less important than the timing of bringing the technology to market. While such a rule of reason approach may not eliminate patent breadth conflicts entirely, litigation will be limited to actions between the cooperative venture or domestic/regional industry members and their global competitors.

Problems associated with incentives for innovation connected with patent breadth also arise with respect to patent length. The broader and longer the patent, the less incentive for innovation beyond the race to be first; i.e. the greater are patent race and monopoly distortion problems. Too short a patent life may not allow for a fair return on investment to innovators, thereby discouraging socially desirable innovation. The shorter the patent length, all other things being equal, the lower the rate of innovation:

> it is desirable to minimize the patent life in order to minimize the duration of monopoly distortions. We should notice, however, that minimizing the patent life could have a distortionary effect not yet discussed: It could reduce the rates of R&D investment . . . if one is concerned both to reduce monopoly distortions and to bolster rates of R&D investment, we presume that the optimal patent life would be such that firms make positive profit.
>
> (Green and Scotchmer, 1993: 12)

Since patent life (defined statutorily) is not a discretionary factor in patent regimes, little in this regard is to be gleaned directly from SEMATECH.

Nevertheless, the behaviour of participating member and non-member firms suggests that patent length (at least in the case of technology-driven cooperative innovation in a global market) is not as significant an incentive factor as commonly believed. As with patent breadth, however, 'it is misleading to ask' how long patent length should be 'without asking simultaneously' whether first mover advantages provide sufficient incentive for innovation. Given the increased rate of innovation, timing and adaptability in respect of manufacturing, marketing and distribution (first-mover advantages) appear to be the more important factors driving the innovative process.

Analysis of their historical rationale and three key issues confronting patent law regimes, in the context of the maturing Age of Information and Global Marketplace as well as in light of the SEMATECH experience, is informative. The discussion has demonstrated that seen beyond their modern context and illuminated by SEMATECH, patent law regimes in Canada and the US are potentially less problematic than conventional wisdom would suggest. At a minimum, this is so with respect to technology-driven innovation[28] because cooperation as competition (globally) renders issues of patent breadth and length less relevant. This is not to say that the analysis and resulting conclusions are universally applicable:

> Since the free riding and scale economies arguments are less pronounced for production activities than for R&D activities, it is far less clear that collaboration to engage in production, distribution and marketing should receive the same special antitrust treatment as has been afforded to cooperative research.
>
> (Shapiro and Willig, 1990: 121)

It is the case, however, that global competition and technology-driven growth in the late 1990s and twenty-first century warrant new ways of approaching patent and related competition law regimes.

RECONCEPTUALIZING COMPETITION: FROM COMPETITION POLICY TO COMPETITIVE INDUSTRIAL POLICY

As conceived since at least the days of Adam Smith, the notion of 'winners and losers' has been paramount to the conceptualization of competition. Accordingly, conventional wisdom would view the results of the SEMATECH experiment to date as most significant with respect to evidence of its contribution to the US semiconductor industry's apparent victory, having re-established global market share leadership. It is important to remember, however, that global competition is dynamic, particularly in the context of globalism dominated by high-technology TNCs:

> Global leadership can shift if industry structural change provides the opportunity for leapfrogging to new products or new technologies that nullify past leaders' scale and learning; . . . the first mover to the new [generation of products or] . . . technology often wins.
>
> (Porter, 1986: 37)

According to the conventional mindset, SEMATECH-like cooperative industrial policy (industry-driven, true consortium, etc.) might be seen as an important component in achieving or preventing such leapfrogging becoming or remaining victorious.

This discussion suggests, however, that viewing competition from so narrow a perspective may itself be the source of the perceived crisis in Canadian and American competitiveness. Heading into the twenty-first century, competition must be understood in all of its dimensions. As such, competition is a more complex and dynamic, multidimensional phenomenon, taking on different forms, depending on time, place and circumstances:

> The intensity or effectiveness of one form of competition often correlates negatively with the strength of other forms. This means that maximization of competition is a meaningless goal. The goal is more correctly described as choosing a preferred mixture of competitive forms.
>
> (Demsetz, 1992: 207)

SEMATECH presents such New Competition in respect of its so-called cooperation in the 'pre-competitive' stage.

To call the cooperation embodied by SEMATECH 'pre-competitive', as if its activities are not part of the 'real' competitive process/market is a misnomer generated by those viewing the enterprise from the two-dimensional 'winners and losers' perspective.[29] As prior discussion demonstrates, SEMATECH's activities encompass R&D, process and product innovation and development, quality control, information and technology transfer; all guided by information and parameters emanating from coordinated market research and strategic planning. Except for the fact that SEMATECH, Inc. is a non-profit corporation[30] and does not market processes or products under its own brand name, other than being frozen in an outdated conception of competition, there ceases to be any basis for the distinction between so-called 'pre-competition' and competition itself.

There are numerous elements of the dimensions of this New Competition, including 'winning and losing' differently understood. Competition is premised on staying in the game, win or lose:

> Cooperation and moral community coexist in a reciprocal and self-reinforcing relationship. To create a consortium in the semiconductor industry, leaders had to behave as members of a community with the superordinate goal of preserving their industry [i.e. staying in the game].
>
> (Browning *et al.*, 1995: 128)

The global competitive imperative is no longer, principally, a matter of winning (of being Number One, the market share leader). Rather, it is the need to become/remain a viable (profitable) participant in the global market place, aspiring at appropriate times for the leadership position but not basing government policy design and implementation,[31] first and foremost, on winning *per se*.

Just as significantly, the elements of dimensionality are not fixed, but potentially as fluid as the movement of various forms of capital, products and services, not to mention changing forms of government. Today and into the twenty-first century, in addition to time (the silent fourth dimension),[32] the key elements of the New Competition germane to competition policy are: global, regional, and domestic markets often, but not necessarily, interrelated (Dimension 1); supply of liquid, fixed and human capital, goods and services (including information and technology), and TNCs (Dimension 2); and, global, regional and domestic demand for consumer, producer, quasi-public and public goods and services (Dimension 3).

The elements of each dimensional component are familiar from the modern competition perspective. In addition, it must be kept in mind that elements within a given 'dimension' can interact, just as the dimensions themselves are interrelated. Consider the element of TNCs, a principal focus of this chapter. Through its modern neoclassical conceptualization, competition between TNCs is understood readily; or is it? The conventional view is that competition between TNCs is acceptable, unless in competing they happen to cooperate or 'collude'. SEMATECH teaches, however, that where the interests of home-based TNCs and domestic interests coincide, cooperation *qua* collusion may be necessary. While such cooperation may 'substantially lessen competition'[33] at certain stages in the continuum of bringing products or services to market (particularly in respect of the domestic market alone), such cooperation can be essential to domestic industries (TNCs and SMEs) being globally competitive. Moreover, such cooperation can have significant and desirable spillover effects:

> In such [technology-intensive] industries, costs fall and product quality improves as the scale of production increases, the returns to technological advance create beneficial spillovers for other economic activities, and barriers to entry generate market structures rife with first-mover advantages and strategic behavior. *A nation's competitive position in industries with these characteristics is less a function of strategic interactions between its factor endowments and more a function of strategic interactions between its firms and government, and between them and the firms and governments of other nations.*
>
> (Tyson, 1993: 3; emphasis added)

From the modern neoclassical perspective, including its conception of competition, given large enough spillover effects, cooperation which 'substantially lessens competition' may be justified as necessary in that it stems from market failure. The New Competition suggests, however, that spillover effects and market failure correction are added benefits, but not the underlying rationale for such cooperation. Instead, due to the reality of competition between governments and nations/regions for the location of TNCs, such cooperation may be fundamental to global competitiveness:

> In today's global economy, in which international sourcing and markets are as important as, if not more important than, their domestic equivalents, the configuration of TNC activity (particularly within the Triad) is less dependent upon the availability and cost of unimproved resources, and more upon the knowledge base and infrastructure facilities of economies in which they are producing or contemplating producing. *At the same time, Governments can and do strongly influence the extent, quality and cost of those factors by their education, science and technology, industrial, trade, environmental, transport and communications and fiscal policies. Indeed, in a variety of ways, nation-sates are increasingly competing for resources and capabilities offered by TNCs.*
>
> (Dunning, February 1992: 162–3; emphasis added)

Given competition for TNCs between governments and nations/regions, from a competition policy perspective, it is essential to consider factors which determine a government or country/region's competitiveness globally:

> A country is a desirable global platform in an industry if it provides an environment yielding firms domiciled in that country an advantage in competing globally in that particular industry . . . the second determinant of the attractiveness of a country's global platform in an industry are the characteristics of a country's demand and local operating environment.
>
> (Porter, 1986: 39)

Porter (1990) further develops this view of the constituent attributes of nations as competitive global platforms, enunciating his 'diamond' paradigm. The diamond consists of four attributes: (1) quantity and quality of demand for goods and services by a country's consumers; (2) level and composition of its natural resources and created factor capabilities; (3) domestic rivalry of wealth-producing agencies (i.e. nature and extent of inter-firm competition); and, (4) extent to which firms are able to benefit from agglomerative or external economies by being spatially grouped in clusters of related activities. Notice that Porter does not view government policy (except to the extent policy leads to governments playing the role of 'wealth-producing agencies') as an attribute of the diamond. Instead, consistent with the modern neoclassical view, he views government as an exogenous factor which shapes and influences the diamond. For example, in explaining Japanese competitive advantage in the 1980s, Porter 'points to the larger number of firms in Japan than in the United States which are competing with each other, as a factor making for the higher industrial competitiveness of the former country' (Dunning, February 1992: 155).

Many of the Japanese firms Porter refers to, however, are interconnected via *keiretsu*, which domestic competition policy has not only permitted but also encouraged. Thus, the sheer number of firms may not be indicative of alleged increased domestic rivalry. Moreover, such analysis ignores the importance of other factors. These include lack of foreign competition within Japan due to lack of access to the Japanese domestic market, which would tend to increase the number of domestic players.[34] Such domestic limits on foreign competition also enhanced the opportunity to use US-originated technology to develop manufacturing at sufficient scale, initially for the domestic market and subsequently, with improvements thereby facilitated, for export. In so doing, Japan achieved improved global competitiveness, at times characterized by significant competitive advantages.

A third factor Porter ignores (mentioned in the Introduction) is the fact that, since the Second World War, Japanese economic development has been restricted to commercial industrialization. Given that military procurement throughout the Cold War (part of government policy) generated industries characterized by few rather than many participants (understandable, at the very least, from a national security perspective), the absence of military industrial development may also account for the relatively greater number of firms in Japan.

Just as important, perhaps most significant, is Porter's failure to consider the difference between US and Japanese government relations with industry:

> McFetridge points out that the rapid growth of the Japanese and German economies in the post-World War II period was achieved without the benefit of vigorous competition law enforcement. On this basis, he suggests that causality may run from national competitive advantage to strong domestic rivalry rather than in the reverse direction (as Porter asserts).
>
> (Anderson and Khosla, 1993: 14)

> In Japan . . . this distinction [between government and industry] is much more nuanced, leading many observers to conclude that the discretionary bureaucratic power and big-business power are impossible to disentangle. The invisible hand is at work in Japan, but it is not Adam Smith's invisible hand – it is the invisible hand of the government working with Japanese industry.
>
> (Tyson, 1993: 57)

Therefore, Porter's diamond paradigm is incomplete. Its resulting myopia is due, principally, to its modern neoclassical foundation. The New Competition necessitates the addition of government as an attribute of national/regional competitiveness internationally/globally:

> in a very real sense, national administrations compete with each other to ensure that their macroeconomic and organizational strategies and policies are such as to provide their wealth creators with the maximum possible incentives to sustain and advance their competitiveness, vis-a-vis [*sic*] their foreign rivals.
>
> (Dunning, December 1992: 21)

Government policy itself becomes an endogenous factor in the Information Age platform (reconceptualization) of global competition.

A second policy-related dimension that emerges with increased significance is the interrelationship between domestic and global competition:

> In most national capitals and in the general public there is no real understanding that the new international arena is within the border. Separating trade policy from competition policy, R&D, financial market regulation and so forth only made sense when tariffs were the most important obstacle to international linkage.
>
> (Ostry, 1992: 22)

In other words, the Dimensional components and elements of the New Competition specified above will at times interact simultaneously. This interaction has serious implications (not captured in Porter's 'diamond' paradigm) beyond the fact that government policies in general, as well as competition policy specifically, become elements of effective New Competition policy.

The most important implication in the context of this discussion relates directly to competition policies prevailing in Canada and the US:

> Just as TNCs are developing new organizational structures to exploit the advantages of global markets and production to counter the measures of their foreign competitors, so, too, Governments need to re-examine their own cultures of governance to take account of the same phenomena.
>
> (Dunning, December 1992: 34)

> Neither Japan nor Europe share the dominant American view that policies to foster specific high-technology industries are at best unnecessary and at worst counter productive. Nor do they share

> American views on competition policy, R&D policy, and intellectual property protection, all of which have important consequences for global competition.
>
> (Tyson, 1993: 45)

This means that government needs to change its view of what constitutes competition. In short, Canadian and US public sector policy planning and implementation must be founded on internalization of the specified dimensions and dimensional elements of the New Competition.

The classical notion of unregulated (no government influence) and freely competitive markets (prevalent once again in modern neoclassical thinking), where the invisible hand generates winners and losers with resulting efficiency, is an ideal ill-suited to the economic realities of the late 1990s and twenty-first century:

> Too much competition and too little cooperation can cause intolerable inequities and instability . . . The doctrine of laissez-faire capitalism holds that the common good is best served by the uninhibited pursuit of self-interest. Unless it is tempered by the recognition of a common interest [cooperation] that ought to take precedence over particular interests [the traditional competitive imperative], our present system – which, however imperfect, qualifies as an open society – is liable to break down . . . there is something wrong with making the survival of the fittest a guiding principle of civilized society. This social Darwinism is based on an outmoded theory of evolution, just as the equilibrium theory in economics is taking its cue from Newtonian physics. The main point I want to make is that cooperation is as much a part of the system as competition, and the slogan 'survival of the fittest' distorts this fact . . . We have now had 200 years of experience with the Age of Reason, and as reasonable people we ought to recognize that reason has its limitations. The time is ripe for developing a conceptual framework based on our fallibility. Where reason has failed, fallibility may yet succeed.
>
> (Soros, 1997: 48–58)

In other words, in the context of global competition in the twenty-first century, notwithstanding the unprecedented levels and pace of technology-based innovation, the Information Age is characterized by the fact of imperfect information ('fallibility'). Such fallibility is best addressed in the spirit of competition as cooperation. SEMATECH demonstrates that cooperation, and perhaps other forms of behaviour heretofore deemed anti-competitive, are key ingredients to competitiveness on the maturing global playing field.

Thus, for example, given increasing global competition between TNCs, less domestic rivalry or even domestic market dominance may cease to be so significant a factor as current competition policy dictates:

> Moderate to strong competition from imports is an important means of improving the allocation of resources and their use. It is also an effective way of curbing the exercise of market power, particularly when production technology calls for scales typical of natural monopolies or when a few dominant producers are entrenched and protected by high entry barriers.
>
> (Frischtak, 1992: 75)

> in the right circumstances, rivalry by foreign firms might offer greater [competitive] benefits than that between domestic firms.
>
> (Lawrence, 1987: 157)

> in discussing the implications of internationalization for antitrust, Charles F. Rule, then U.S. Assistant Attorney General, has remarked, 'Effective foreign competition serves as a more expedient and efficient check on competitive abuses by domestic firms than U.S. antitrust enforcers can ever hope to be.'
>
> (Anderson and Khosla, 1993: 12)

Moreover, this analysis suggests that in the case of high-technology industries and TNCs, with respect to being competitive globally, domestic cooperation may be more important than rivalry between domestically based TNCs:

> Industrial policy advocates such as Lester Thurrow have suggested that competition policy may deter necessary business restructuring and joint entrepreneurial efforts, and even questioned the need for competition policy in the presence of increasing global competition.
>
> (Anderson and Khosla, 1993: 2–3)

> Japan has attained superior structural competitiveness despite acute resource deficiencies, relative political isolation among the industrialized democracies, proximity to the large unstable transnational economies in Russia and China, and increasing strains in trade with the USA. The structural competitiveness has been achieved through excellence in applied technology, *through the efficiencies of intercorporate cooperation, and through very functional forms of technocratic-managerial collaboration.*
>
> (Boyd, 1996: 209; emphasis added)

> Social networks in Asia operate to support economic growth, productivity and competitiveness by reducing the 'social distance' of relationships and the antagonisms of Western pluralism, in favour of more cooperative relations among individuals and between them and the authorities . . .
>
> (Crone, 1996: 9)

Indeed, to 'force' domestic rivalry through competition policy measures may even be counterproductive to maintaining or improving a nation's global competitiveness.

Since the focus of this chapter is on high-technology TNCs, one final illustration of the policy implications of the New Competition paradigm is instructive. Recall in the discussion of SEMATECH the emphasis on its approach to intellectual property and evolving treatment of patents for innovations developed. The treatment, moving away from exclusive rights and towards 'common' market rights, calls for reconsideration of the patent component of competition policy. It suggests that the global competitive imperative renders restrictions on licensing, on the one hand, and patent length, on the other, inappropriate. This is particularly the case where product cycles are short and first-mover advantages great, as is typically the case in high-technology industries.

Elaboration of the conceptualization of the New Competition, prompted by recognition of the role of TNCs and examination of SEMATECH, indicates the need for systematic changes in the approach to competition policy:

> despite the recent trend toward economic 'liberalization' around the world, many nations remain committed to the goal of supporting their high-technology producers. Certainly this is the case in the European Community, in Japan, and in many newly industrializing countries like Korea, Taiwan, and Brazil. Most of the high-technology industries targeted and nurtured by these nations are ones in which the United States has considerable competitive strength and which are important to the US industrial base in a variety of ways.
>
> (Tyson, 1993: 4)

In Canada and the US, without substantive changes in government policy, industries in which global competitive advantage exists today face increased risk of becoming less competitive tomorrow:

> If we are to stay a competitive force, it is essential that we seize the opportunities and build a strong economy on a platform of

> technological leadership. It involves business, government and universities working in a new spirit of cooperation.
>
> (Fisher, 1993: 19)

SEMATECH provides an enlightening display of such cooperation, cooperation which has a potentially broad array of beneficial effects:[35]

> cooperation [which] requires a common vision that spans the entire semiconductor industry (including its suppliers and customers) and extends as well to academia, government, and electronics users. The pervasive importance of semiconductor-based electronics – coupled with the high cost, complexity, and sophistication of IC [integrated circuit] technology of the late 1990s – compel the creation of this new shared vision. Success requires expanded teamwork and cooperation throughout the private and public domestic semiconductor infrastructure. It must be vision driven by customer and systems needs and not by technology alone.
>
> (Burger and Howard, 1993: 23)

> What is novel and theoretically interesting about SEMATECH, then, is that it offers insight into how cooperation can arise and persist in a highly competitive industry.
>
> (Browning *et al.*, 1995: 114)

Such vision must be shared by government, as it is called upon to participate in the reconceptualization of competition. Moreover, government must incorporate the New Competition in its development and implementation of substantively new forms of policy. The global competitive imperative of the twenty-first century demands replacement of current competition policy with CIP which includes substantive revisions of relevant aspects of today's competition policy regimes.

COMPETITION REGIMES OF CANADA AND THE UNITED STATES: FRIEND OR FOE TO GLOBAL COMPETITIVENESS?

The question raised in the above heading may seem rhetorical. It is important, however, to apply the foregoing analysis to specific aspects of prevailing competition regimes in Canada and the United States.[36] Such application serves three purposes. First, it demonstrates that existing policy is amenable to critical review on the basis of the New Competition; i.e. the reconceptualization of competition is grounded in much of modern neoclassical reality. Second, it develops further understanding of the New Competition paradigm. Third, it suggests ways in which competition policy might be better formulated (anything but cursory consideration of which is beyond the scope of this chapter):

> Recognizing . . . [the] often negative correlations between different dimensions of competition, I take as the goal of antitrust [competition policy] the achievement of an efficient mixture of competitive forms or its mirror image an efficient mixture of monopolistic forms.
>
> (Demsetz, 1992: 207)

The complex dimensionality of the New Competition is intrinsic to the analysis that follows.

In the context of high-technology TNCs, given the natural tendency towards domestic concentration,[37] the focus of the discussion is on merger and collusion aspects of current competition policy regimes.[38] Canadian merger policy is legislated under the Competition Act (amended, 1985, 1990):

> Although the merger provisions in the Canadian Competition Act have yet to receive much authoritative interpretation, the following general principles can be stated:
>
> - Competition concerns are paramount – *there is little or no role for industrial or social policy objectives.*
> - Although structural considerations are not irrelevant, *competition is treated primarily as a behavioural process of rivalry between firms which is desirable in order to prevent accumulations of market power that would result in higher prices or reduced choice for consumers.*
> - Transactions which would facilitate the exercise of market power either unilaterally (i.e. by a dominant firm) or interdependently (i.e. by an oligopoly) are proscribed.
>
> (Campbell and Trebilcock, 1992: 9; emphasis added)

Thus, Canada's merger regime, derived directly from the modern neoclassical conceptualization of competition, is diametrically opposed to the kind of regime that would facilitate effective CIP.

The merger regime in the US is only marginally better:

> Since General Dynamics, merging parties in the United States faced with prima facie cases based on concentration data have successfully rebutted the presumption of anti-competitiveness by

> presenting evidence regarding such factors as ease of entry, efficiencies and changing market conditions.
>
> (Campbell and Trebilcock, 1992: 19)

While US courts have to some extent recognized the global competitive imperative, like in the Canadian regime, the onus in merger disputes remains with the merging parties. A regime which recognizes the New Competition, with its attendant need for and value of CIP, would take a different approach. The onus would be on plaintiffs to prove the anti-global competitiveness of the transaction, taking into account the dimensionality and relevant elements of the New Competition.

In respect of collusion, the Canadian approach views such activity as criminal. Collusion is seen as so contrary to social values inherent in the prevailing conceptualization of competition that the behaviour warrants criminal liability, both corporately and personally for directors. On its face, the Competition Act allows for defences based on some of the key components of CIP:

> Section 45 [the conspiracy provisions] explicitly allows for defenses in the areas of R&D joint ventures, product standards, advertising, bulk buying, the dissemination of technical information, and export consortia as long as such arrangements do not have an undue effect on competition in the Canadian market. However, despite these defenses and the clarifications provided by the PANS decision, it is worth considering whether the conspiracy provisions could give rise to a chilling effect on the establishment of strategic alliances in Canada.
>
> (Competition Policy Bureau, 1993: 16)

In theory, the PANS decision provides for a 'partial rule of reason' approach.[39] 'Since the provisions were adopted, [however,] there has not been a single case in which a specialization agreement has been registered or a joint venture been approved on the basis of the provisions' (Anderson and Khosla, 1993: 32). The facts are that a strong chilling effect is already in place.

It is fair to say, therefore, that Canadian competition policy and its US counterpart, to the extent they place the onus for cooperation defences or collusion exceptions on the parties themselves, are not CIP-friendly:[40]

> [Canadian and] U.S. policy must decide whether prevention of collusion, and thus any sharing of post-development marketing data, should be the ultimate goal, or whether emphasis should be shifted in favor of encouraging all aspects of innovation. The fact that successfully restraining competition through collusion in the

> current complex and fast-paced technological market is highly unlikely leans heavily in favor of the latter. In short, U.S. antitrust [and Canadian competition] policy must fully embrace innovation as a primary goal in order for any allowable cooperative efforts to maximize the economic benefits of technological development.
> (Sendra, 1992: 388)

Consideration of the merger and collusion components of Canadian and US competition policy demonstrates that each fails to take into account the global imperatives of the New Competition on which the need for CIP are based.

Further clarification of the New Competition readily results from its application to specific competition policy provisions. In the case of special agreements referred to above (relative to the collusion provisions), the New Competition approach entails designing competition policy in light of the policy of competing jurisdictions:

> This [the special agreements situation] is in contrast to the experience in other jurisdictions that provide special treatment for such arrangements. For example, in the European Community, specialization agreements have been used to facilitate restructuring in light of the progressive unification of the European market.
> (Anderson and Khosla, 1993: 32)

Japan and the EC are ahead of Canada and the US in their treatment of special agreements:

> the EEC clearly leads in the area by providing companies and joint ventures with a 'safe harbor.' Rather than being subject to Commission approval, joint ventures which involve high investment risk or that produce important new technology can bypass 85(1) completely, and be directly exempted under the 85(3) block exemption (if they do not extend beyond the limits set by the black list). In contrast, under the U.S. National Cooperative Research Act, joint ventures are still subject to rule of reason analysis and the accompanying threat of damages.
> (Sendra, 1992: 390)

In Japan, as has already been pointed out, where competition policy would apply it is not enforced. As in Europe, exemptions are readily given to high-technology special agreements.

For mergers the situation is no different than that of special agreements. In the EC, cross-border mergers among TNCs increased more than 300 per cent between 1987 and 1990, and 'the number of such transborder acquisitions in 1990 for the first time exceeded the number of such acquisitions involving firms in a single member country' (Vernón, 1992: 20). The integrated nature of Japanese industry previously identified has similar consequences. Canadian and US policy-makers must take into account the industrial restructuring dynamics of their global competitors:

> Old notions of competition, monopoly, and patent power will have to be redefined in terms of the current requirements of innovation. Larger social goals such as technological and economic progress will have to be clearly integrated into the language of our legal regimes.
>
> (Sendra, 1992: 407)

Application of the New Competition paradigm to prevailing competition policy in Canada and the US highlights the need to utilize a reconceptualization of competition:

> In the European Community [and Japan], the role of competition policy is extensively intertwined with industrial policy and the objectives of European unification . . . Even in the U.S., with perhaps the strongest tradition of independent enforcement agencies and a highly legalized antitrust system, industrial policy concerns arguably are being given increasing attention in competition policy implementation.
>
> (Anderson and Khosla, 1993: 25)

> specialization agreements which bring together companies in different member states are promoted and are less likely to be challenged under the Community's competition policy rules. E.C. rules in this regard could be particularly relevant to the future application of competition policy rules under the NAFTA award.
>
> (Competition Policy Bureau, 1993: 22)

From the Canadian perspective, elements of competition within North America *vis-à-vis* the US and Mexico, and globally, with respect to the US, Mexico and others, make the need for reconceptualization all the more compelling.

Therefore, for both Canada and the US, incorporating the New Competition in the design and implementation of competition policy is necessary:

> It is important to ensure that competition law and policy in Canada are compatible and competitive with corresponding regimes of other major industrialized countries. In the global economy of the 1990s, the design of economic framework policies and

> institutions has itself become an important means by which jurisdictions compete with each other for access to the scarce technology, capital and skilled labour inputs [typically controlled by TNCs] which are essential to continuing prosperity.
>
> (Anderson and Khosla, 1993: 5)

In significant circumstances, notwithstanding domestic competition concerns *per se*, concentration within a given country or trading bloc may be vital to achieving global competitiveness relative to TNCs based in a competing nation or region like Japan or the EC. CIP, which facilitates this and other changes in Canadian and US competition policy, is dictated by the global competitive imperative and better understood using a reconceptualization of competition as the New Competition.

The discussion is left, then, to consider changes to Canadian and US competition policy. First, of course, is the need to acknowledge the need for and acceptance of a reconceptualization of competition. It has already been suggested that in respect of both mergers and collusion, not only is illegality *per se* contrary to the New Competition, but CIP calls for going beyond a rule of reason approach which places the onus on the allegedly breaching parties. Effective CIP, as embodied in SEMATECH, requires, minimally, a rule of reason approach which reverses the onus. This has been recognized judicially in the US:

> [In *Continental T.V., Inc.* v. *GTE Sylvania Inc.*, 433 U.S. 36 (1977), 'facial unreasonableness'] replaces notions of per se illegality with the recognition that even anti-competitive practices may have favorable aspects and should be analyzed in an economic context on a case by case basis.
>
> (Sendra, 1992: 386)

This analysis indicates that such an approach needs to go further even than that adopted in the enabling legislation for SEMATECH:

> [The National Cooperative Research Act] covers all activities considered essential to bringing an idea 'close to' technological commercial application, such as exchanging research data, setting up facilities and applying for patents. *The NCRA, however, specifically excludes production, marketing, and even joint research focused on preparing the product for the marketplace.*
>
> (Sendra, 1992: 387; emphasis added)

In addition, effective CIP must embody a similar reverse onus/rule of reason approach in respect of specialization agreements.

The New Competition recognizes that exclusions under the NCRA are artificial and counter productive, motivated by policy design and implementation based on the modern neoclassical view of competition:

> [Canada and] the United States still has [have] not created a mechanism that recognizes that a competition policy must not only restrain anti-competitive practices but must also create or even adapt such practices to stimulate production and foster innovation.
>
> (Sendra, 1992: 392)

The first step required in order to create such a mechanism is adoption of a new paradigm based on the reconceptualization of competition:

> If one hundred years of federal antitrust (competition) policy has taught us anything, it is that . . . [such policy] is both political and cyclical. Almost every political generation has abandoned the policy of its predecessors in favor of something new.
>
> (Hovenkamp, 1985: 213)

The New Competition goes even further, calling for competition policy which takes into account 'political' and other global realities, elements of the dimensional components which define a new conception of competition. With the global competitive imperative dictating the need for CIP derived from and consistent with the New Competition, the time has come for 'something new'.

CONCLUSION

The foregoing discussion of TNCs as the dominant actors in globalization, the emergence of relationships between TNCs and governments based on mutually beneficial negotiation, the case of SEMATECH as an example of such a relationship and a viable model for CIP incorporating the New Competition, implications of SEMATECH for prevailing patent law regimes, and elaboration of the New Competition paradigm and its application to key aspects of current Canadian and US competition policy reveal a complex drama on the global stage. Nevertheless, clear and present lessons for competition policy emerge.

First, is the need to acknowledge and adopt a reconceptualization of competition, as is done in the New Competition paradigm. Second, is the need to design and implement competition policy which facilitates industrial policy initiatives; that is, the adoption of CIP in place of conventional competition policy. For example, contrary to views critical of existing patent law regimes, the SEMATECH experience suggests that intellectual property regimes in Canada and the US generally suffice. One exception to their adequacy is the fact that the statutory definition of patent length and limitations on licensing may be unnecessary, even at times counter-productive, to innovation and growth.

In the future, cooperation between TNCs and governments embodied in CIP may lead to new competition problems globally, analogous to the anti-trust problems that emerged towards the end of the nineteenth century:

> Over time, the differences that heretofore have distinguished United States-based from European- and Asian-based trans-national networks are likely to diminish, as the conditions of their founding and early growth begin to lose their original importance.
> (Vernon, 1992: 21)

The final lesson derived from this discussion is that before a new age of 'anti-competitiveness' arrives, governments are well-advised to work together, just as TNCs are doing, in 'competition' with the natural and predictable tendency of emerging 'global' TNCs towards globally 'anti-competitive' behaviour:[41]

> Where control of some sort still seems necessary or desirable, the option remaining will be to pursue mutually agreed measures with other countries. In the decades ahead, the United States, Europe and Japan are sure to find themselves addressing the feasibility and desirability of international agreements that define more fully the rights and obligations of transnational corporations.
> (Vernon, 1992: 27–8)

Just as the New Competition views relationships between TNCs and governments being based upon mutually beneficial negotiation, so it engenders the need to forge ahead with relationships between national and regional governments based upon mutually beneficial negotiation. From this perspective, the completion of the Uruguay Round of GATT negotiations (albeit, with a very narrow consensus) indicates that governments may be opening their eyes in time to design and implement a global competition regime which will facilitate prevention of the extreme adjustment problems of the recent past and present; problems, in no small measure, generated by New Competition dynamics operating upon conventional, outdated industrial/competition policy regimes.

NOTES

1 '[D]riven primarily by Cold War objectives, the US established a security regime in the region (Asia Pacific, and Japan in particular) that stabilized politics, allowed the domestic emphasis to shift from military to economic spending, and fostered an economic environment that was, broadly speaking, liberal in its trade and investment policies' (Crone, 1996: 4). Accordingly, perhaps it is not coincidental that US and Canadian re-emergence in the mid-1990s as leading edge global competitors in manufacturing has followed the end of the Cold War.

2 The majority of TNCs are technology-driven, whether in their core products or services themselves, or in the use of high technology in the information processing or manufacture/delivery of non-technological products or services.

3 '[T]he nineteenth century neoclassical paradigm derived from Newtonian physics may be changed for the new scientific approach based not on an immutable universe, but on a world in perpetual change, where equilibrium [rather than the be all of competitive markets] is only a dimension of dynamic systems' (Niosi, 1996: 99).

'Economic efficiency may be the result of cooperation, as well as competition, and a comparative analysis of both types of arm's-length transactions may be worth-while' (Niosi, 1996: 103).

4 A recent example of high-technology TNC merger activity is the February 1997 US$6.6 billion merger between Netscape and US Robotics.

5 The role of transnational corporations . . . in the world economy has increased dramatically in the past decade or so, and this trend is expected to continue. For instance, the stock of world outward direct investment increased almost fourfold from US$519 billion in 1980 to US$2 trillion in 1992, considerably outpacing the growth of world output and trade (Rao and Legault, 1996: S475).

6 See Dunning (December 1992: 11).

7 The words 'competition and anti-trust' in the context of statutory policy are used interchangeably in this chapter.

8 The non-global era of the nineteenth and twentieth centuries was characterized by a profound lack of free trade. Prior GATT efforts notwithstanding, it is only since the end of the Cold War, with the establishment of the World Trade Organization and completion of the Uruguay Round (1994), that a global era characterized by free trade has become possible, if not yet a reality.

9 See Kennedy (1992: 67).

10 See Dunning (December 1992: 34).

11 Information not attributed to other sources has been obtained through confidential interviews with SEMATECH personnel during August, September and November 1993, and May 1997.

12 See GAO (1992: 32). '[T]he matching funding from the federal government was also a necessary, if not sufficient, input. Although the semiconductor firms did not really want government involvement, early planning indicated that they could not afford to commit all the revenues required (Browning *et al.*, 1995: 143).

13 See SEMATECH (1990: 6).

14 SEMATECH clearly demonstrates that in the field of technology-related inter-firm (and public institution) agreements, 'one is in fact moving from the analytical ground of *unplanned* spillovers and related externalities to that of linkages formed in at least a partially conscious manner with the purpose of reaping

partly predefined or *targeted returns* to technological activity through *cooperation*, e.g. in a collective manner' (Chesnais, 1996: 21).

15 As of 1996, 100 per cent of annual funding is provided by member companies. In other words, other than its enabling legislation, SEMATECH, Inc. has been privatized.

16 'The lesson learned in the first four years of operation include: improved ways for companies to work together; methods of cooperation between defense programs and industry programs; the ability to set objectives and missions of interest both to the defense and commercial market sectors; ways to bring engineers and scientists from member companies to work in teams to develop new equipment and pre-competitive manufacturing methods; effective ways to disseminate information to member companies and the government; and the development of an industry-wide Quality program' (Spencer and Grindley, 1993: 26).

17 See Tyson (1993: 41–2).

18 As modern neoclassical economists, the Chicago School and others might want to claim.

19 ROI formulas are complicated, and another benefit that has emerged has been the development of rational, data-driven R&D investment decisions based on a ROI formula developed at SEMATECH.

20 'Every way I look at the data, I come back to the belief that this has got to be a SEMATECH-driven increase [in world market share for semiconductor equipment made in America]' (Jerry Hutcheson, Chief Executive Officer, VLSI Research Inc. As quoted in the *San Francisco Chronicle*, 9 April 1992, in SEMATECH, Inc., *1992 Annual Report*, 1993: 11).

21 'SEMATECH has helped give us much more than mere percentage points on a market share chart. SEMATECH has given our industry – and our country – a model for the future' (Robert W. Glavin, Chairman, Executive Committee, Motorola, Inc., in SEMATECH, Inc., *1992 Annual Report*, 1993: 23). The success of SEMATECH is further evidenced by the fact 'that 11 of the original 14 member companies and the U.S. government agreed to extend their membership in SEMATECH for a second five-year period and committed themselves to new and expanded goals. Their continued commitment to SEMATECH indicated that they believed that SEMATECH had achieved something worth their investments. As Craig Barnet of Intel told us, "I judge SEMATECH by results. The organization set out to recover market share from Japan; five years later, market share has been recovered. At Intel we call that a results-oriented, successful project"' (Browning *et al.*, 1995: 119).

22 See GAO (1992: 16).

23 See SEMATECH (1990: 7). In addition, member firms are given a 6–9 month lead in preparing for and integrating new technologies into their manufacturing process: 'Several member companies have incorporated aspects of SEMATECH's facility design in their own wafer fabs. One member company has based the entire design of a large, new fab on SEMATECH's facility' (SEMATECH, 1990: 7).

24 Similarly, the notion of too much competition being counterproductive has also evolved.

25 See Scotchmer (1991: 35).

26 For example, Northern Telecom (NorTel, Canada's largest electronics firm) 'has been an associate and a shareholder [since 1990] of the Microelectronics and Computers Technology Corporation (MCC) . . . MCC is a privately owned and financed non-profit cooperative joint venture of some twenty large

American (plus NorTel) corporations . . . Founded in 1982 . . . MCC conducts collaborative research on a large number of areas of computer electronics, including advanced computing technology, CAD, software technology and interconnection . . . It is not an industry–university consortium, nor is it officially linked to any government, though it has government sponsors including DARPA, NASA and the US Department of Defense. (Niosi, 1996: 112–13).

27 See Merges and Nelson (1993: 24).

28 Whenever innovation is subject to competition between TNCs, as well as between nations and international/regional trading blocs, patent breadth and length become less relevant, if not altogether irrelevant.

29 'A strong case can indeed be made for introducing into growth theory an expression of the resource-creating, value-creating and surplus-creating potentialities which stem from cooperation – in particular when the existence of and satisfactory functioning of networks, involving many firms and directed towards technological innovation and/or the production of highly specialized manufacturing or business services, can be established' (Chesnais, 1996: 22).

30 A question of interest beyond the scope of this chapter is whether, given the performance of SEMATECH and its subsequent privatization, such cooperative ventures in the future can/should be private, for-profit corporations from the outset. Given the CIP advocated in this chapter, special enabling legislation would not be needed, so the formation of such ventures could be left altogether to the private sector.

31 Neither should private (TNC) strategic planning and implementation be based first and foremost on winning *per se* (being/becoming the biggest market player). For example, gaining market share through merger does not seem to have produced expected efficiency enhancing results: 'The literature suggests that while most ex ante studies predict merger-induced efficiency gains, the vast majority of mergers do not seem to produce the predicted efficiencies. Indeed, the literature suggests that mergers seem to fail 60–80% of the time' (Hartman, 1996: 296).

32 A fundamental difference between the modern neoclassical conception of competition and that of the New Competition paradigm (which includes time as a fourth dimension) is that the former is based on application of Newtonian mechanics, whereas the latter is derived from application of modern quantum physics.

33 The 'substantial lessening of competition' is the threshold standard (legal test) for prosecution under the Canadian Competition Act.

34 'It is in this latter area – international problems that demand international solutions – that the United States has much to learn from the basic economic blueprint for the EC internal market, which tore down frontier barriers; conceptualized competition, trade, and national industrial policy issues as a seamless web; and prohibits significant unjustified public, private, and hybrid restraints' (Fox, 1996: 727)

35 Competition as cooperation has a synergistic effect unavailable to competitors independently: 'Once the resource- and surplus-creating capacity of firms has been established, one can start looking at the similar potential that networks might possess: *organizations which possess the capacity to create resources can take steps to ally their strengths and increase their capacity through cooperation* [emphasis added]. On the basis of the linkages they can establish among themselves (as well as with innovation-related public institutions) through a wide variety of contractual and informal arrangements, firms can combine and pool innovative and productive capabilities which if properly coordinated can

be greater than the sum of the technological knowledge and capabilities of the individual participant firms and institutions (Chesnais, 1996: 28).

In addition, such cooperation can generate positive returns beyond those typically reflected in bottom line accounting terms, but which ultimately impact economic performance: '[M]any of the benefits of cooperation, at least to an organization, can be defined in noneconomic terms; benefits might include fast cycle time of product to market, improved quality, high-quality decision making, improved competitiveness, and so on. These dimensions can be seen as the intervening variables that help to explain why cooperation might enhance performance and satisfaction' (Smith *et al.*, 1995: 17).

36 Decreased access to the Japanese domestic market is a result of restrictive government FDI rules and the inhibiting impact of *keiretsu* on FDI.

37 Scale and barrier to entry factors, as well as the proclivity for reorganization through merger, joint ventures and strategic alliances, account for the natural tendency towards domestic concentration.

38 Given the author's experience, and space limitations, the discussion is centred on Canadian competition policy.

39 See Anderson and Khosla (1993: 30).

40 To require special legislation authorizing cooperative ventures as in the case of SEMATECH is also an example of merger (competition) policy being CIP-unfriendly.

41 Signs of it emerging, for example, may be increased incidence of cross-border mergers between leading TNCs.

BIBLIOGRAPHY

Anderson, R. D. and Khosla, S. D., 'Competition Policy as a Dimension of Industrial Policy: A Comparative Perspective' (Ottawa: Policy Directorate, Bureau of Competition Policy, Consumer and Corporate Affairs Canada, 22 January 1993).

Bhagwati, J., *The World Trading System at Risk* (Princeton: Princeton University Press, 1991).

Blaine, M. J., 'America's Structural Competitiveness', in G. Boyd, ed., *Structural Competitiveness in the Pacific: Corporate and State Rivalries* (Cheltenham, UK and Brookvield, VT: Edward Elgar, 1996).

Boyd, G., 'Japan's Structural Competitiveness', in G. Boyd, ed., *Structural Competitiveness in the Pacific: Corporate and State Rivalries* (Cheltenham, UK and Brookvield, VT: Edward Elgar, 1996).

Brodley, J. F., 'Antitrust Law and Innovation Cooperation', *Journal of Economic Perspectives*, Vol. 4, No. 3, Summer 1990.

Browning, L. D., Beyer, J. M. and Shetler, J. C., 'Building Cooperation in a Competitive Industry: SEMATECH and the Semiconductor Industry', *Academy of Management Journal*, Vol. 38, No. 1, 1995.

Burger, R. and Howard, W., *Semiconductor Technology: SIA Workshop Conclusions* (San Jose, CA: Semiconductor Industry Association, 1993).

Campbell, N. and Trebilcock, M., 'A Comparative Analysis of Merger Law: Canada, The United States and the European Community', *World Competition*, March 1992.

Chandler, A. D., Jr, 'The Evolution of Modern Global Competition', in M. E. Porter, ed., *Competition in Global Industries* (Cambridge, MA: Harvard University Press, 1986).

Chesnais, F., 'Technological Agreements, Networks and Selected Issues in Economic Theory', in R. Coombs *et al.*, eds, *Technological Collaboration: The Dynamics of Cooperation in Industrial Innovation* (Cheltenham, UK and Brookvield, VT: Edward Elgar, 1996).

Comanor, W. *et al.*, 'Competition Policy in a Global Economy', Unpublished draft (Toronto: University of Toronto–Centre for International Studies, 1992).

Competition Policy Bureau, *Strategic Alliances and Their Implications for Canadian Competition Policy* (Ottawa: Competition Policy Bureau, March 1993).

Coombs, R., Richards A., Saviotti, P. P. and Walsh, V., 'Introduction', in R. Coombs *et al.*, eds, *Technological Collaboration: The Dynamics of Cooperation in Industrial Innovation* (Cheltenham, UK and Brookvield, VT: Edward Elgar, 1996).

Crone, D., 'The Pacific in the World Economy', in G. Boyd, ed., *Structural Competitiveness in the Pacific: Corporate and State Rivalries* (Cheltenham, UK and Brookvield, VT: Edward Elgar, 1996).

Demsetz, H., 'How Many Cheers for Antitrust's 100 Years?', *Economic Inquiry*, April 1992.

Dodgson, M., 'Learning, Trust and Inter-Firm Linkages: Some Theoretical Associations', in R. Coombs *et al.*, eds, *Technological Collaboration: The Dynamics of Cooperation in Industrial Innovation* (Cheltenham, UK and Brookvield, VT: Edward Elgar, 1996).

Doz, Y. L., 'Government Policies and Global Industries', in M. E. Porter, ed., *Competition in Global Industries* (Cambridge, MA: Harvard University Press, 1986).

Dunning, J. H., 'The Competitive Advantage of Countries and the Activities of Transnational Corporations', *Transnational Corporations*, Vol. 1, No. 1, February 1992.

Dunning, J. H., 'The Global Economy, Domestic Governance, Strategies and Transnational Corporations: Interactions and Policy Implications', *Transnational Corporations*, Vol. 1, No. 3, December 1992.

Fisher, G., 'Sharpening the US Edge', *Wordlink*, July/August 1993.

Fox, E. M., 'Antitrust, Competitiveness, and the World Arena: Efficiencies and Failing Firms in Perspective', *Antitrust Law Journal*, Vol. 64, 1996.

Frischtak, C. R., 'From Monopoly to Rivalry: Policies to Realize the Competitive Potential of Transnational Corporations', *Transnational Corporations*, Vol. 1, No. 2, August 1992.

Gandolfo, G., *International Economics I: The Pure Theory of International Trade* (Berlin: Springer-Verlag, 1987).

GAO (Resources, Community, and Economic Development Division), 'SEMATECH'S Technological Progress and Proposed R&D Program', GAO Briefing Report to the Chairman, Subcommittee on Defense, Committee on Appropriations, US Senate (Washington, DC: United States General Accounting Office, July 1992).

Ghemawat, P. and Spence, M. A., 'Modelling Global Competition', in M. E. Porter, ed., *Competition in Global Industries* (Cambridge, MA: Harvard University Press, 1986).

Gold, B., 'Technology, Markets and Politics: Closing the Defense Technology Gap Between America and Europe', American Political Science Association Annual Meeting, Washington, DC, 1991.

Gray, H.P., 'John Dunning's Eclectic Paradigm', *Transnational Corporations*, Vol. 1, No. 1, February 1992.

Green, J. and Scotchmer, S., 'On the Division of Profit in Sequential Innovation', in S. Scotchmer, ed., *Economics of Intellectual Property Casebook* (University of Toronto: Faculty of Law, 1993).

Grindley, P., Mowery, D. C. and Silverman, B., 'SEMATECH and Collaborative Research: Lessons in the Design of High-Technology Consortia', *Journal of Policy Analysis and Management*, Vol. 13, No. 4, 1994.

Grossman, G. M. and Helpman, E., *Innovation and Growth in the Global Economy* (Cambridge, MA: MIT Press, 1993).

Hansson, G., *Harmonization and International Trade* (New York: Routledge, 1990).

Hartman, R. S., 'Predicting the Efficiency Effects of Mergers', *Journal of Forensic Economics*, Vol. IX, No. 3, Autumn 1996.

Hovencamp, H., 'Antitrust Policy After Chicago', *MLR*, Vol. 84, November 1985.

Irwin, D. A., 'The US in a New Global Economy: A Century's Perspective', *American Economic Review*, Vol. 86, No. 2, May 1996.

Jorde, T. M. and Teece, D. J., 'Innovation and Cooperation: Implications for Competition and Antitrust', *Journal of Economic Perspectives*, Vol. 4, No. 3, Summer 1990.

Kennedy, C. R., Jr, 'Relations Between Transnational Corporations and Governments of Host Countries: A Look to the Future', *Transnational Corporations*, Vol. 1, No. 1, February 1992.

Kenworthy, L., 'Balancing Competition and Cooperation', *Challenge*, July–August 1996.

Krugman, P., 'Competitiveness: A Dangerous Obsession', *Foreign Affairs*, March–April 1994.

Lawrence, P. R., 'Competition: A Renewed Focus for Industrial Policy', in D. J. Teece, ed., *The Competition Challenge* (Cambridge, MA: Ballinger, 1987).

Martin, P. 'Global Industries and the Political Economy of U.S. Response to Foreign Industrial Targeting', American Political Science Association Annual Meeting, Washington, DC, 1991).

Merges, R. and Nelson, R., 'On Limiting or Encouraging Rivalry in Technical Progress: The Effect of Patent Scope Decisions', in S. Scotchmer, ed., *Economics of Intellectual Property Casebook* (University of Toronto: Faculty of Law, 1993).

Milner, H. V., *Resisting Protectionism: Global Industries and the Politics of International Trade* (Princeton: Princeton University Press, 1988).

Niosi, J., 'Strategic Technological Collaboration in Canadian Industry: Towards a Theory of Flexible or Collective Innovation', in R. Coombs *et al.*, eds, *Technological Collaboration: The Dynamics of Cooperation in Industrial Innovation* (Cheltenham, UK and Brookvield, VT: Edward Elgar, 1996).

Ohmae, K., 'Death of the Nation State', *Worldlink*, July/August 1993.

Ostry, S., *Governments & Corporations in a Shrinking World: Trade & Innovation Policies in the United States, Europe and Japan* (New York: Council on Foreign Relations Press, 1990).

Ostry, S., 'The Domestic Domain: The New International Policy Arena', *Transnational Corporations*, Vol. ?, No. 1, February 1992.

Porter, M. E., 'Competition in Global Industries: A Conceptual Framework', in M. E. Porter, ed., *Competition in Global Industries* (Cambridge, MA: Harvard University Press, 1986).

Porter, M. E., *The Competitive Advantages of Nations* (New York: The Free Press, 1990).

Rao, S. and Legault, M., 'An Overview of Canadian Direct Investment Abroad', *Canadian Journal of Economics*, Special Issue, April 1996.

Rowley, W. J. and Baker, D. I., eds, *International Mergers: The Antitrust Process* (London: Sweet & Maxwell, 1991).

Rugman, A. *et al.*, *International Business; Firma and Environment* (New York: McGraw-Hill Book Co., 1985).

Scherer, F. M., 'Nordhaus' Theory of Optimal Patent Life: A Geometric Reinterpretation', *American Economic Review*, 1972.

Scherer, F. M., 'Industrial Market Structure and Economic Performance', in M. Handler *et al.*, eds, *Patents and Antitrust* (New York: Foundation Press, 1983).

Schmalensee, R., 'Antitrust and the New Industrial Economics', *American Economic Review*, May 1982.

Scotchmer, S., 'Standing on the Shoulders of Giants: Cumulative Research and the Patent Law', *Journal of Economic Perspectives*, Winter 1991.

SEMATECH, Inc., 'High-Technology Research Consortia: SEMATECH'S Experience' (Austin, TX: SEMATECH, Inc., 1990).

SEMATECH, Inc., *1992 Annual Report* (Austin, TX: SEMATECH, Inc., 1993).

SEMATECH, Inc., *SEMATECH Update* (Austin, TX: SEMATECH, Inc., January 1993).

SEMATECH, Inc., *SEMATECH Update* (Austin, TX: SEMATECH, Inc., July/August 1993).

Sendra, M., 'Strategic Alliances for Innovation in the Global Market of the 1990s: A Comparative Study of the Relationship Between Innovation and the Patent/Antitrust Mechanisms of the United States and the European Economic Community', *International Tax and Business Lawyer*, Vol. 9, No. 2, Winter 1992.

Shapiro, C. and Willig, R. D., 'On the Antitrust Treatment of Production Joint Ventures', *Journal of Economic Perspectives*, Vol. 4, No. 3, Summer 1990.

Smith, K. G., Carroll, S. J. and Ashford, S. J., 'Intra- and Interorganizational Cooperation: Toward a Research Agenda', *Academy of Management Journal*, Vol. 38, No. 1, 1995.

Soros, G., 'The Capitalist Threat', *Atlantic Monthly*, February 1997.

Spencer, W. J. and Grindley, P., 'SEMATECH After Five Years: High-Technology Consortia and U.S. Competitiveness', *California Management Review*, Summer 1993.

Spencer, W. J. and Tobias, P. S., 'Statistics in the Semiconductor Industry: A Competitive Necessity', *The American Statistician*, Vol. 49, No. 3, August 1995.

Thurrow, L. C., *Head to Head: The Coming Economic Battle Among Japan, Europe and America* (New York: William Morrow, 1992).

Trebilcock, M., 'The Evolution of Competition Policy: A Comparative Perspective', in M. Mathewson, M. Trebilcock and M. Walker, eds, *The Law and Economics of Competition Policy* (Vancouver: The Fraser Institute, 1990).

Tyson, L., *Whose Bashing Whom: Trade Conflict in High Technology Industries* (Washington, DC: Institute for International Economics, 1993).

Vernon, R., *Storm Over the Multinationals: The Real Issues* (Cambridge, MA: Harvard University Press, 1977).

Vernon, R., 'Transnational Corporations: Where Are They Coming From, Where Are They Headed?', *Transnational Corporations*, Vol. 1, No. 2, August 1992.

Wallace, D., Jr, *International Regulation of Multinational Corporations* (New York: Praeger Publishers, 1976).

11

COMMENTARY

Derek Ireland

The chapters by Donald McFetridge and Richard Braudo address many of the same issues that I and my former employer, the Bureau of Competition Policy, have been wrestling with for the past number of years. In my case, with much greater depth, sophistication and precision than I have been able to bring to these questions. If I have one quibble with the authors, it is with the use of the term 'industrial policy' in a positive manner. Based on my 27 years of experience in industrial policy and related fields, I have a bad habit of equating industrial policy in Canada with 'rent seeking behaviour', and not any of the other more positive attributes ascribed to it by, for example, Richard Braudo. In fact, I was under the impression at one point that it was a term that had fallen out of favour.

In my own work, I had reached the point where I was using other terms, like 'innovation policy', 'competitiveness policy', even 'competition policy', instead of 'industrial policy'. While recognizing that these were less precise terms, I am almost sorry to see industrial policy-making a comeback in Canada. I am sure this reflects some kind of defect in my character, and not an integral problem with industrial policy, but one still wonders if the comeback of industrial policy is progress.

Beyond this quibble, I found myself agreeing with most of what is discussed in the two chapters. I would argue, however, that in practice, if not in law, competition policy in Canada, the US and most of the other major antitrust jurisdictions can find and has found ways to accommodate the cooperative arrangements, strategic alliances, innovation and dynamic change that are the subjects of this book.

I would also argue that for many, this accommodation takes place in ways that are awkward, uncertain, arbitrary, less than fully transparent and perhaps, at times (for some business actors), even menacing. There are many reasons why I have raised this concern, including the following:

1 In a number of jurisdictions, strategic alliances and other innovative arrangements will be addressed under criminal, *per se* type prohibitions. Even with all the qualifiers, exemptions and exceptions

employed by all of us (including the partial rule of reason approach in Canada), the argument can and has been made (albeit without strong empirical support) that this has a chilling effect on innovation in Canada and elsewhere. At the very least, it is odd that mergers, which are forever and involve all of the operations of the merging parties, can receive more permissive treatment under competition law than a strategic alliance, which involves only a portion of the operations of the parties and may last only for a short period of time.

2 In Canada, and perhaps elsewhere, it is not clear under which part of the competition law an innovation-based alliance will be treated. The Bureau's recent draft information bulletin on the topic lists no less than six separate sections which could be relevant to strategic alliances, ranging from the traditional criminal conspiracy provisions to the merger provisions with their efficiency gains defence.

3 Supporters of Canadian competition policy and law, including myself, like to point to the flexible, efficiency based approach of much of the current statute, and to the many innovation-enhancing provisions of the law: the undueness test and the R&D exemption in the conspiracy provisions; the superior competitiveness performance factor in the abuse of dominance provisions; the efficiency gains defence, and the foreign competition and other factors in the merger provisions; the specialization agreement and joint venture provisions; the treatment of intellectual property, and so on.

To date, these and other provisions have been applied by the Director in a pragmatic, even accommodating manner. Richard Braudo can see menace in our merger provisions, but up to now only a handful of mergers have been challenged by the Director before the Tribunal, and in the two most important challenges the Tribunal ruled in favour of the parties. Unless I have missed something, a Canadian court has never ruled against a merger in over 106 years. Canadian merger law can be more readily criticized as overly permissive and a paper tiger, than as a threat to pro-competitive mergers and acquisitions.

4 Still, there remains a problem. For most of these accommodating provisions in the Canadian law there is almost no jurisprudence. It is argued by some that the accommodation has been largely through the application of administrative discretion by the Director. As Donald McFetridge stresses, we can only speculate on how the Tribunal or the courts will address these issues in an actual case. If we have any hints on this, they are less than encouraging – such as the views expressed by the Chairman of the Tribunal on the efficiency gains defence in rendering the decision on a merger case a few years ago. Critics of US jurisprudence have also raised concerns about the ability of the American courts to address complex economic issues, like dynamic efficiencies, spillovers, externalities and appropriability.

5 There can also be concerns, particularly for an export-dependent country like Canada, that an arrangement that is deemed to be pro-competitive in one jurisdiction can raise competition concerns in a second. Recent amendments to US law make this a real possibility for Canadian strategic alliances. The danger of private harassment-type suits in US courts adds to the menace. Indeed, critics of the European Commission decision in the De Havilland case see similar concerns across the ocean for innovative business arrangements based in Canada which are breaking into the European market.

In short, I would suggest that the administration of Canadian competition law over the past nine years has been consistent with the spirit of the concerns raised in the Braudo and McFetridge chapters.

The accommodation taking place currently is based on a mix of the law itself, the administrative discretion of the Director and, perhaps, the Bureau's diminishing resources. Therefore, to some degree, this accommodation is subject to change based, perhaps, on a new government or Minister less sensitive to issues of innovation and efficiency enhancing corporate arrangements; or even, to some degree, a new Director who from the perspective of the authors would be more doctrinaire and less pragmatic.

Despite the substantial modernization which took place a decade ago, the Competition Act still offers some leeway for an overzealous Director to cause considerable dismay in the boardrooms of the nation, and perhaps some harm to the efficient development of the Canadian economy: the RPM provision, for one, comes to mind; and, we also have refusal to supply just to add to potential alarm. A number of academic experts (including Professor Michael Trebilcock) have offered modest suggestions to address these problems through amendments to the Competition Act, particularly with respect to the treatment of strategic alliances and other potentially pro-competitive arrangements which do not fall under the merger provisions. There has yet to arise, however, a groundswell of support from the business community to reopen the Competition Act and address these concerns. (In fact, among the Bureau's more recent public consultation efforts, the consultation on confidentiality raised a lot more interest and debate among business people and their legal advisers than the one of strategic alliances.)

When it comes to strategic alliances and the treatment of innovation, intellectual property and dynamic efficiencies under competition law, business people and politicians alike may require evidence of real harm, not merely speculation on a possible chilling effect. Until such harm is readily apparent, this issue is not going to rank up there with the debt, Quebec's independence, medicare and the other great public policy issues of the day – nor should it.

Finally, there are a few other points I would like to make:

1 I find myself questioning, from both a competition policy and industrial policy perspective, the relevance of the semiconductor and SEMATECH experience to a small open economy like Canada, which is dependent on imported technologies, foreign investment and secure access to foreign markets. How would a successful Canadian R&D consortium like SEMATECH be treated under US trade and competition laws, and in the American courts? Is there, in fact, a level playing field?

Our ongoing trade disputes suggest that a positive answer to this question could be naïve. Some of our Japanese friends could feel the same way, given the recent twists and turns in their trade disputes with our American friends. And would the head offices of foreign subsidiaries in Canada allow them to participate in a Canadian SEMATECH?

I appreciate the point that SEMATECH was industry-driven. Otherwise, in my view, it would not have worked. Who are the industry drivers of similar arrangements in Canada? I, for one, refuse to believe that the Canadian Competition Act is a reason for the almost total lack of these kinds of high profile arrangements in this country.

2 Competition law is a general law of general application, and should be applied equally to all sectors. Competition rules should be able to accommodate innovation and technological change. However, should they be 'modernized' to fit only the high technology sectors?

The same can be said of intellectual property. I know that semiconductors and SEMATECH are important. However, how relevant are their experiences and needs to pharmaceuticals, chemicals and the entertainment industry, which have been at the forefront of the IP debate in North America, were the major catalysts behind the TRIPs negotiation at the GATT, and have also raised competition issues from time to time?

3 To what extent should competition policy and law be driven by a relatively few high-technology global industries. What about those more mature, perhaps less interesting industries where technologies and products do not change daily (e.g. cement, iron and steel, automobiles, aluminium, airlines, etc.)? These industries are not very exciting, but they are still important to national economies, international trade and consumer spending.

If we were to modernize our competition law so that it fits high technology like a glove, what would it mean for all those industries which for one reason or another continue to be oriented towards local and provincial markets, and are protected from the forces of the international marketplace and the 'New Economic Paradigm'?

4 In short, could we be arguing for a different set of competition rules for exciting, strategic, high technology industries? Is this possible, or desirable? Or, is this going to lead to a new form of rent-seeking behaviour, where every industry lines up to be designated 'strategic'?

INDEX